Natural Resource Valuation and Policy in Brazil

METHODS AND CASES IN CONSERVATION SCIENCE
Mary C. Pearl, Editor

METHODS AND CASES IN CONSERVATION SCIENCE

Thomas K. Rudel and Bruce Horowitz
Tropical Deforestation: Small Farmers and Land Clearing in the Ecuadorian Amazon

Joel Berger and Carol Cunningham
Bison: Mating and Conservation in Small Populations

Jonathan D. Ballou, Michael Gilpin, and Thomas J. Foose
Population Management for Survival and Recovery: Analytical Methods and Strategies in Small Population Conservation

Susan K. Jacobson
Conserving Wildlife: International Education and Communication Approaches

David S. Wilkie and John T. Finn
Remote Sensing Imagery for Natural Resources Management: A First Time User's Guide

Gordon MacMillan
At the End of the Rainbow? Gold, Land, and People in the Brazilian Amazon

PERSPECTIVES IN BIOLOGICAL DIVERSITY SERIES

Holmes Rolston III
Conserving Natural Value

SERIES EDITOR
MARY C. PEARL

SERIES ADVISERS
CHRISTINE PADOCH AND DOUGLAS DALY

Natural Resource Valuation and Policy in Brazil: Methods and Cases

Edited by Peter H. May

COLUMBIA UNIVERSITY PRESS
NEW YORK
1999

Columbia University Press
Publishers Since 1893
New York Chichester, West Sussex

Library of Congress Cataloging-in-Publication Data

Natural resource valuation and policy in Brazil : methods and cases /
edited by Peter H. May.
p. cm. — (Methods and cases in conservation science)
Includes bibliographical references.
ISBN 0-231-10826-5 (cloth : alk. paper). — ISBN 0-231-10827-3
(pbk. : alk. paper)
1. Natural resources—Valuation—Brazil. 2. Natural resources—
Government policy—Brazil. I. May, Peter Herman. II. Series.
HC187.5.N38 1999
333.7′0981—dc21

99-25861

Casebound editions of Columbia University Press books
are printed on permanent and durable acid-free paper.
Printed in the United States of America
c 10 9 8 7 6 5 4 3 2 1
p 10 9 8 7 6 5 4 3 2 1

Contents

Acknowledgments

The chapters in this volume resulted from a pioneering Ecological Economics project in Brazil (Project Eco-Eco), with support from the Rockefeller Foundation. This project was initially conceived shortly after UNCED (the United Nations Conference on Environment and Development)—the Earth Summit—in Rio de Janeiro in 1992, when it was perceived that the nation had few scientists trained in the use of integrative ecological and economic principles and techniques that would be essential to the identification and assessment of sustainable development proposals.

As part of its activities, Project Eco-Eco carried out a workshop in September 1993 in São Paulo, uniting sixty Brazilian and international specialists in this area to debate the methods used and the implications of the papers presented. A Portuguese version of the series was then published (May 1995), which has been widely read by Brazilian scholars and policy analysts and has now become a fundamental reference in training and in discussion of issues in resource valuation and conservation policy.

Two additional case studies and a theoretical/methodological chapter (chapter 1) have been added to the collection for the current English version. The latter seeks to familiarize the non-Brazilian reader with the policy setting for natural resource development and management there. The author acknowledges support from the Brazilian National Council of Scientific Research (CNPq) toward preparation of this chapter during his sabbatical research at the University of California-Berkeley. The material in this volume has not previously been published in English, with the exception of chapter 3 by Almeida and Uhl

(1995) and prior versions of chapter 2 (May 1994) and chapter 8 (May and Pastuk 1996).

CHAPTER ACKNOWLEDGMENTS

Chapter 2 was initially prepared for presentation at the International Expert Meeting on Sustainable Resource Management and Resource Use in Noordwijk, Netherlands, 3–4 June 1993, and it was subsequently published (May 1994). The analysis was based in part on results of the research project "Estimates of Environmental Accounts for Brazil" by the Institute for Applied Economic Research (IPEA-Rio, Ministry of Planning, Brazil), with support from the United Nations Development Program. The recent data on Amazon and Atlantic Forest depletion were presented by the author at the Workshop on "National Environmental Accounting: A Practical Guide," World Bank, March 1996.

The research on which chapter 3 is based was made possible through financial support from the Ford Foundation. The authors thank Joshua Bishop, Peter May, Michael Collins, and Carlos Young for help with the economic analysis; Paulo Barreto, Marli Mattos, Angélica Toniolo, and Adalberto Veríssimo for providing raw data and clarification on many points; Eugenio Arima, Robert Buschbacher, Ted Gullison, David McGrath, Roger Moeller, Daniel Nepstad, Campbell Plowden, Harrison Pollak, Eustaquio Reis, Johan Sweede, Ricardo Tarifa, Mariella Uzeda, and Robert Walker for reading earlier versions of this paper and providing many helpful suggestions; João Flor and Lucia Porpina from SECTAM, and Conceição Silva e Renato Coral from SAGRI for special help with data acquisition; and, finally, we thank all the people in the public and private agencies in Belém (SEPLAN, EMBRAPA, SEFA, SETRAN, DER, BASA, and SAGRI) and Paragominas (BASA, EMATER, the Rural Workers' Union, and the municipal government) for their time and understanding.

Much of the material in chapter 4 was derived from dissertation research by the author, Josemar X. de Medeiros, who expresses his appreciation to Professor Oswaldo Sevá, of the State University of Campinas in São Paulo, for his penetrating insights on the consequences of energy resource choices and alternatives.

The authors of chapter 5 thank Dr. Antonio Evaldo Comune of the Faculty of Economics and Administration of the University of São Paulo for the courage to explore a new, polemical, but necessary field in Brazil—the economic valuation of mangroves.

The authors of chapter 6 wish to acknowledge the kind cooperation of the Companhia Independente de Polícia Florestal do Estado do Mato Grosso do Sul (CIPFlo/MS), and the British Council (Brasília) for the academic exchange funding of both authors for natural resource valuation in Brazil's Pantanal region. Helpful comments were provided by Carlos E. F. Young. The Centre for Socio-Economic Research on the Global Environment (CSERGE) is a designated research center of the Economic and Social Research Council, United Kingdom.

Chapter 7 is based on research conducted at the Institute for Applied Economic Research (IPEA), a branch of Brazil's Planning Ministry, with assistance from the Graduate Program in Development, Agriculture and Society of the Federal Rural University of Rio de Janeiro (CPDA/UFRRJ). The studies received partial financial support from the National Council for Scientific Research (CNPq) (project no. 521524/94) and the International Center for Economic Growth (ICEG). The latter is a nonprofit international policy institute that has contributed to economic growth and human development in developing and transitional countries by enhancing the capacity of indigenous policy research institutes to foster the policies and institutions of a market economy. To accomplish this, ICEG sponsors a wide range of programs—including research, publications, conferences, seminars, and special projects advising governments—through a network of 370 member institutes worldwide.

For the case studies presented in chapter 7, Claudio S. de Sousa and Carolina Dubeux assisted in data collection and tabulation for the first case, while Edgar Menezes and Renata Soares provided assistance for the second. The authors are grateful to Peter May for his comments and review.

Chapter 8 arose from a didactic exercise at CPDA/UFRRJ used in training the growing number of professionals and graduate students in Brazil eager to apply concepts and tools of ecological economics, counteracting the absence of case study materials available in Portuguese related to the urban environment. Original research involving fieldwork in São José Operário was an integral part of training; the authors acknowledge the contributions of twelve course participants during the initial fieldwork. Further on-site fieldwork, institutional interviews, and benefits estimation were undertaken by an interdisciplinary team including, besides the author (a natural resource economist), an environmental sociologist, and specialists in watershed management, forestry, soil science, biology, and political science.

The author thanks the staff of the Community Reforestation Project of the municipal government of Rio de Janeiro, particularly Celso J. Santos and Márcia Garrido, for logistical assistance and project documentation. More important than this, the author thanks the residents of São José Operário themselves, and particularly their Residents' Association, for agreeing to participate in the study, and for assisting in the administration of questionnaires. Finally, support from the Jessie Smith Noyes Foundation, the Rockefeller Foundation, and the Conservation, Food and Health Foundation is gratefully acknowledged, as are valuable comments made on a previous version by Karl Steininger and Marc Dourojeanni. Appreciation is expressed to the editors of Island Press and the International Society for Ecological Economics for permitting the article to be republished in slightly different form. For any errors that remain, the author assumes the full burden of responsibility.

Peter H. May
Berkeley, December 1998

Contributors

Oriana Trindade de Almeida is affiliated with the Instituto de Homem e Meio Ambiente da Amazônia (IMAZON). She has been instrumental in consolidating several years of innovative ecological and socioeconomic research by this innovative institution, whose objective is to offer opportunities for young professionals in the applied sciences, agronomy, and forestry, to contribute to policy formation on land-use change.

Monica Grasso, a doctoral candidate in coastal and estuarine environment studies at the University of Maryland, is associated with the Maryland International Institute for Ecological Economics. She is a marine biologist experienced with mangrove assessment, environmental valuation, and ecosystem modeling, having received her master's degree at the University of São Paulo Oceanographic Institute.

Peter H. May is Professor of Agrofood Systems and Sustainable Development at the Federal Rural University of Rio de Janeiro, Graduate Program in Development, Agriculture and Society. He previously served as Program Officer for Rural Poverty and Resources at the Ford Foundation, and Forest Officer (Nonwood Forest Products) at the United Nations Food and Agriculture Organization in Rome. Founder and past president of the Brazilian Society for Ecological Economics, he is a member of the editorial board of *Ecological Economics* and coeditor of *Pricing the Planet: Economic Analysis for Sustainable Development* (Columbia University Press 1996). He was a Visiting Scholar at the Energy and Resources Group, University of California-Berkeley in 1998.

Josemar Xavier de Medeiros is Professor of Energy Systems Planning at the University of Brasília and has until recently served as Project Analyst at the National Council for Science and Technology Development (CNPq).

Francisco Eduardo Mendes, an oceanographer, is a candidate for the doctorate in Energy and Environmental Planning at the Graduate Engineering Program of the Federal University of Rio de Janeiro, having served as a Research Associate in the Environmental Studies section of the Research Institute for Applied Economics-IPEA of Brazil's Ministry of Planning.

André Steffens Moraes is a socioeconomist at the Center for Agricultural and Livestock Research for the Pantanal, Brazilian Corporation for Agricultural and Livestock Research (EMBRAPA-CPAP), Corumbá, Mato Grosso do Sul.

Dominic Moran is a research associate at the Centre for Social and Economic Research on the Global Environment (CSERGE) at University College London and the University of East Anglia, UK. He is currently serving as an advisor to the Government of Malaysia.

Leonardo Rezende is an economist in the master's program at the Federal University of Rio de Janeiro Institute of Economics.

Yara Schaeffer-Novelli is Professor of Marine Biology at the Oceanographic Institute of the University of São Paulo, and former Sub-Secretary for Research of the São Paulo State Secretariat of the Environment.

Ronaldo Serôa da Motta is Senior Researcher and Coordinator of Environmental Studies in the Research Directorate of the Research Institute for Applied Economics (IPEA) of Brazil's Ministry of Planning, in Rio de Janeiro, and is a professor of Environmental Economics at the Universidade Santa Ursula in Rio de Janeiro. He is coeditor of *Pricing the Planet* (Columbia University Press 1996).

Christopher Uhl is Professor of Forest Ecology at Pennsylvania State University and a Pew Fellow. He is cofounder and scientific advisor of IMAZON in Belém, Pará.

1

Globalization, Economic Valuation, and Natural Resource Policies in Brazil

PETER H. MAY

The widespread exhaustion and degradation of natural resources in Brazil today exhibit characteristics similar to those found in many developing nations. Although Brazil is an emerging nation with a modern industrial sector and sophisticated commercial agriculture, poverty persists for the majority of the nation's population of over 160 million. Over much of Brazil's hinterland, marginalized rural communities dependent on natural resources for their survival coexist with large enterprises that extract these resources as raw materials for the production of consumer goods and commodities for an ever more globalized market. Both types of actors are responsible in some measure for the degradation that has occurred. The former are victims of an unequal society that has often expelled them from lands they formerly cultivated, forcing them to occupy fragile frontier territories. The latter, whose actions obey the logic of international capital, intensify their activities in search of ever greater profits.

This process is by no means new. The history of Brazil is marked by cycles whose economic flows are strongly linked to patterns of land and resource use. A deeply rooted perception of a society with unbounded horizons, based on the sheer immensity of its territory and untapped natural wealth, created a false sense of optimism regarding perspectives for growth and satisfaction of the society's needs. Nevertheless, constraints on effective access to natural resources by the majority of the rural population continue to restrict their efficient and sustainable use.

HISTORICAL PATTERNS OF RESOURCE USE IN BRAZIL

The patterns of spatial occupation in Brazil were the result of production cycles of the goods that constituted the principal export base in distinct historical moments: dyewood, sugarcane, gold, coffee, rubber, and, most recently, soybeans. Initially, these enterprises exhibited an expansion path toward the extensive resource frontier, with growth in production ensured by the opening up of new areas to cultivation and extractivism. This occupation pattern tended to exhaust soil and forest resources closest to major population centers and ports, expanding in circles of ever wider radius in the search for fertile soils and concentrations of natural wealth once the initial areas had been degraded. Although this mobility corresponded to the relative abundance of land, such expansion became inefficient to the extent that it required the installation of a network of transport and market infrastructure that was both difficult and expensive to construct and maintain. This process was marked by concession of large areas to few owners, leading to rigidity of land tenure, limiting access to productive resources, and impeding the development of land markets. With respect to natural resources, soil productivity was exhausted, native tropical biodiversity was impoverished, and water sources dried up.

The economy's reliance on agricultural and other primary products for its income was severely criticized by members of the Dependency School, who prescribed protectionist import substitution tactics as a remedy for what were perceived to be differential terms of trade between developed and developing nations (e.g., Cardoso and Faleto 1969). This approach was applied rigorously in Brazil, as it was in many of the emerging nations during the post-War years.

Import-substituting industrialization promoted the most significant intensification in pressures on the natural resource base that Brazil had yet experienced, despite its emphasis on urban-industrial growth. This was true because a growing primary products sector was necessary to provide capital for industrial investment. An acceleration of economic growth occurred during the late 1960s and throughout the 1970s, when the military government of this era pushed for diffusion of technologies that would enhance land productivity and made investments in infrastructure that would enhance the velocity of wealth extraction and intersectoral transfer. The key elements of this strategy included (i) incentives toward installation of "backward-linked" industries, particularly those needed for production of tractors and implements, chemical fertilizers, and pesticides; (ii) provi-

FIGURE 1.1 Map of Brazilian Regions, Showing Case Study Sites

sion of subsidized credit linked to adoption of "green revolution" technologies; and (iii) the fortification of agribusiness networks and of industrial enclaves based on forest and mineral raw materials, notably in the eastern Amazon and the savannas of the Center-West (see figure 1.1). These investments were stimulated by the availability of cheap petrodollars in the international financial markets after the price shock of 1974, through which Brazil contracted external debt that rose to an astonishing $115 billion by the early 1980s.

Besides the greater intensification in the utilization of natural resources that accompanied this process, the benefits of these advances tended to be concentrated in the hands of those producers and investors that held greater bargaining power with the agents of credit, incentive concession, and markets, having been restricted for the most part to crops and raw materials for export. Besides the inequities associated with this model, the intensification of production systems also led to negative environmental effects, including high erosion rates, water contamination, sedimentation of rivers, chemical intoxication, desertification, and salinization. Finally, the debt crisis of the early 1980s, which led Brazil and the remainder of Latin America to suffer a "lost decade" for human development in the ensuing period, was in part the result of the magnanimous policies of the "miracle" years of the 1970s.

The debt crisis, economic stagnation, and the galloping inflation of the 1980s provoked a dramatic turnaround in recent emphases in Brazilian political economy, in a manner similar to that experienced by other Latin American nations. These changes include a diminished role for the State in the definition of development paths, the privatization of state enterprise, the opening up to foreign investment of various sectors previously reserved for domestic firms, a drastic reduction in import tariffs, transport costs, and other trade barriers, and the formation of regional trade blocks.

These "neoliberal reforms" have raised alarm over the possible effects of these processes on food security and the environment. Recent analyses have assessed the "environmental content" of massive new net inflows of private foreign investment (Gentry et al. 1998), the impacts of liberalization in agricultural trade on land use intensity and labor displacement (May and Segura Bonilla 1997), and prospects for adoption of environmentally friendly production practices (May 1995).

Through these studies, in which the author participated, it has grown clear that the models of development and of natural resource utilization in Brazil have brought environmental costs of great magnitude. This is true for those models inspired both by the exploitation of the extensive margin pressured by the need to service growing external debt, and by intensification of export-oriented resource industries following practices adopted in northern nations. In evaluating these tendencies, there is an evident need to search for solutions more appropriate to the Brazilian institutional and social context.

The papers in this volume present several recent efforts on the part of a growing nucleus of concerned scholars in Brazil to evaluate natural resource and socioenvironmental conflicts in the light of advances in economic theory and regional management experience. The theoretical backdrop draws upon a review and critique of neoclassical models of rational economic behavior in the face of resource scarcity, whose applicability in developing countries is often questionable. This analysis prospered tardily in Brazil, where study of natural resource economics and policy has been adopted over only the past few years in advanced education. Analytical formulae have in most cases been appropriated from studies in other contexts not necessarily relevant to local conditions.

Although natural resource problems possess characteristics that are common among different nations, the configuration of causal factors, the power and property structure prevailing over the resource base, and the foundations of value construction differ among social groups. Thus attempts to copy institutional formats, incentives, and mechanisms to regulate economic actors have often confronted difficulties. Despite such difficulties, a wealth of socio-institutional initiatives have arisen that articulate often divergent interests in an effort to legitimate models of resource management more firmly grounded in local realities (see Lopes et al. 1996).

ENVIRONMENTAL AND NATURAL RESOURCE POLICY FRAMEWORK

This section describes the evolution and current structure of Brazilian environmental and natural resource policy, in particular its legislation and institutional structure, as a framework for assessment of alternative models and instruments.

Until fairly recently, natural resource policy in Brazil was fragmented across a range of federal ministries, with little coordination. Forest policy was the domain of the Institute of Forest Development (IBDF), a branch of the Ministry of Agriculture, which was responsible both for licencing of forest extraction and for the protection of nature reserves and biological resources. Water resource management was subsumed under the auspices of the Department of Water and Energy in the Ministry of the Interior, which also controlled hydroelectric power generation and irrigation works.

A reconsideration of natural resource management policy was born out of concern for environmental issues stimulated by the United Nations Conference on the Human Environment in Stockholm in 1972. In that forum, the Brazilian delegation assumed the posture that the problems of resource depletion and degradation are associated with the excessive consumption common among wealthy nations and should hence be resolved there, while developing nations should be permitted to grow unfettered by environmental safeguards. Despite this adversarial position, environmental legislation that had been limited previously only to preservation of flora and fauna was extended to protection of water quality and the recuperation of lands degraded by mineral extraction.

In 1973, a Special Secretariat for the Environment (SEMA) was created within the Ministry of the Interior. State agencies with environmental protection responsibilities were also created in this period, including the State Foundation for Engineering and Environment (FEEMA) in Rio de Janeiro, and the Company for Environmental Technology and Sanitation (CETESB) in São Paulo. These agencies had as their principal responsibility to respond to industrial pollution problems, through licencing procedures based on what was later to be extended to a unified national System for Licencing of Polluting Activities (SLAP). Management of natural resources and public lands was retained at the federal level, while land use planning and zoning was confined to the few municipalities that were able and willing to create the institutional capacity to cope with land use conflict.

In 1975, the environmental theme was incorporated within the Brazilian national development plan (II PND), whose identification of critical pollution problem areas conditioned the approval of industrial projects on observance of pollution control norms.

In the 1980s, environmental consciousness in Brazil was intensified when industrial health disasters erupted at Cubatão in São Paulo. The comprehensive National Environmental Policy of 1981 was the culmination of Brazil's legislative response to these concerns. The policy proclaimed the following:

> The National Environmental Policy has as its objective the preservation, improvement and recuperation of environmental quality essential to life, and to assure national socio-economic development, in the interests of national security and protection of the dignity of human life . . .
>
> *Federal Law no. 6.938*

This law provided for decentralization of technical-administrative functions regarding the environment among the different spheres of government, ample dissemination of information to the public regarding environmental issues, and participation in review of projects and related activities.

The 1981 law established a range of instruments for achieving its broad objectives:

- Environmental quality standards
- Environmental zoning
- Assessment of environmental impacts (EIA)
- Licencing and monitoring of currently or potentially polluting activities
- Disciplinary or compensatory penalties against infractions

Furthermore, to ensure a more complete integration and coordination of national environmental policy among federal, state, and local governments, the 1981 law created the National Environmental System (SISNAMA) and the National Environmental Council (CONAMA). The former is made up of government agencies responsible for oversight of policies affecting natural resource use or improvement of environmental quality. CONAMA establishes technical and administrative guidelines for implementing environmental policy. Members of CONAMA include, in addition to government agency officials, presidents of the principal trade unions, industrial and agricultural federations, two nongovernmental environmental assemblies, and five civil organizations representing environmental concerns pertinent to distinct macroregions.

Among environmental policy instruments, the most polemical has been EIA, applicable to a wide range of activities that might result in alterations in physical, chemical, or biological properties of the environment. CONAMA's first resolution, in January 1986, required preparation of an EIA and summary public review report (RIMA) to obtain government licencing for such activities, whether in the public or private domain. Among projects subject to EIA/RIMA were included agropastoral establishments over 1,000 hectares (ha), as well as activities causing significant deforestation, and all significant industrial facilities.

In regulating the federal legislation for impact assessment, CONAMA defined the requirements for carrying out public hearings as a channel for participation by individuals or groups concerned with the decisions to be taken regarding licencing of projects that might significantly modify the human environment. As a result of such hearings, project characteristics might be altered or even an entire project halted. In practice, however, the EIA requirement was often misused, and the reports were given little airing in public forums.

Later in the 1980s, extensive burning in the Amazon provoked international concern over CO_2 emissions and possible global warming. The 1988 Constitution reinforced the perception that environmental protection is an essential corollary of socioeconomic development, obliging all levels of public administration to ensure that such development does not degrade the natural environment. In an important change superseding the 1981 legislation, the Constitution delegated to the states significant authority to legislate regarding environmental questions. In an oblique fashion, the new Constitution also delegated to municipalities the competence to enact local master plans (*Planos Directores*) that would orient spatial growth patterns along environmental lines.

The new constitution was also considerably more severe in its penalties, making corporations and governmental authorities alike subject to sanctions if their actions degrade the environment, and providing for criminal penalties. Liability for claims is placed on the agent that causes environmental damage, rather than requiring that the burden of proof be the responsibility of the damaged parties—an application of the Polluter-Pays principle. The regulatory legislation provides for application of fines and, if technological improvements are not met, plant closings and even imprisonment of infractors. The constitution also provides for class action for the defense of diffuse interests.

In other sections referring to environmental protection, the 1988 Constitution conditions the social function of property on environmental preservation, suggesting that activities that satisfy environmental criteria (among other criteria related to labor relations and productive land use) be exempted from expropriation for land reform purposes.

The 1988 constitution and prior environmental legislation require industries to prevent and correct damages caused by pollution emissions. Environmental licencing by state authorities is authorized if the required EIA/RIMA and environmental standards and norms are observed. The latter refer to emissions of gases, vapors, noise, vibrations, radiation, risk of explosion, fire, leakage and other emergencies, raw material volume and quality, employment generation, operating hours and traffic, patterns of land use, and availability of basic infrastructure.

In Brazilian constitutional law, agricultural projects and plantation forestry are conditioned by the adequate establishment of erosion control measures as preparatory activities, as well as agroecological zoning, use of native varieties, and environmental education. A law of 1989 conditioned production, marketing, and use of pesticides on prior registry with the federal government, and on a determination of the degree to which their production and use may endanger the environment and human health, and on their adequate labeling to prevent misuse.

In 1989, the federal government adopted new measures related to the environment, among which was included the creation of a government agency with broad responsibilities in the conduct of environmental policy, the Brazilian Institute for the Environment and Renewable Natural Resources (IBAMA). This superagency was made up of several existing departments previously scattered among several ministries that had been responsible for ecological reserves, pollution control, environmental education, forestry, natural rubber policy, and fisheries.

Finally, influenced by the public concern aroused by the realization of the United Nations Conference on Environment and Development (UNCED) in Rio de Janeiro in June 1992, the Brazilian government created a Ministry of the Environment (MMA) later that year, to which IBAMA was subordinated. In 1994, specific responsibility for coordinating policy relative to the Amazon region was added to the MMA mandate, and in 1995, with the Cardoso government, responsibility over water resource development was also included within this framework. A rough organizational chart is provided in figure 1.2.

The discovery that, despite considerable efforts at containment, the rate of Amazon deforestation and selective timber harvesting had actually increased since UNCED, led the government to enact a provisional policy package in 1996 that curtailed many timber licences in force, banned exports of mahogany, and expanded the proportion of private lands that were to be retained under forest use from 50 to 80 percent. Catastrophic burning and further revelations of accelerated Amazon destruction in January 1998 were among factors that led the Cardoso administration to push through regulations that would make environmental fines and criminal penalties stick, but these fell short of environmentalists' call for wider enforcement powers.

Effectiveness of Policy Enforcement

Brazil's environmental and natural resource policies, widely considered among the most advanced legislative frameworks in the world, have in practice often been difficult to implement. This difficulty is due principally to weak intergovernmental coordination and a lack of definition of responsibilities among federal, state, and municipal

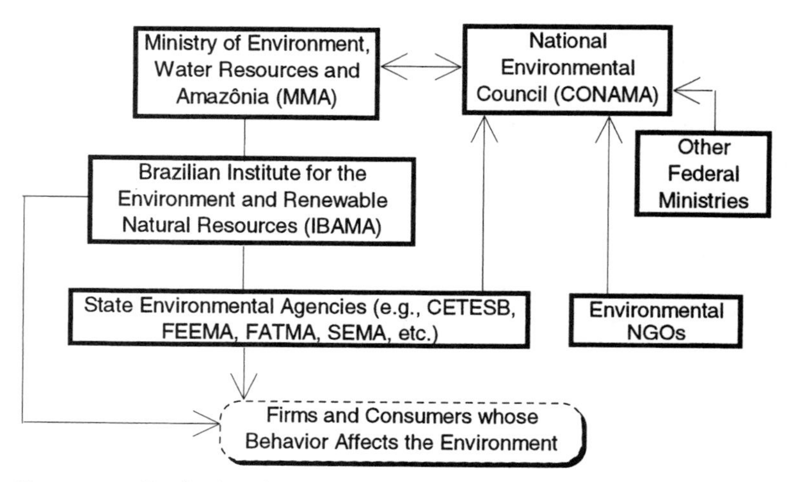

FIGURE 1.2 Institutional Arrangements of Natural Resource and Environmental Administration in Brazil

authorities, as well as to the chronic shortage of funds. A decentralization of functions to state and municipal authorities or nongovernmental organizations is rendered particularly difficult when such decentralization implies the transfer of federal funds and responsibilities, both jealously guarded by IBAMA.

Since environmental restrictions interfere in sectors whose overall regulatory responsibilities fall in the domain of different ministries, such interference is often rejected because of political pressures from stakeholders. It has thus been difficult to integrate environmental concerns in the formulation of macroeconomic or sectoral policies. This has led to a number of policy inconsistencies: among the most glaring of these has been the greater tax rates applied on rural land not used for productive purposes, which has acted as an incentive toward deforestation.[1]

The regulations that were adopted have only rarely been observed in practice because of their precarious enforcement capacity, in turn a function of insufficiency and misallocation of environmentally related taxes and charges that are not directly tied to enforcement-agency expenditures. The establishment of IBAMA as a superagency with an overall purview of natural resource concerns did not resolve these problems, since the formerly dispersed units of which the agency was composed only formed equally distant fiefdoms within the same

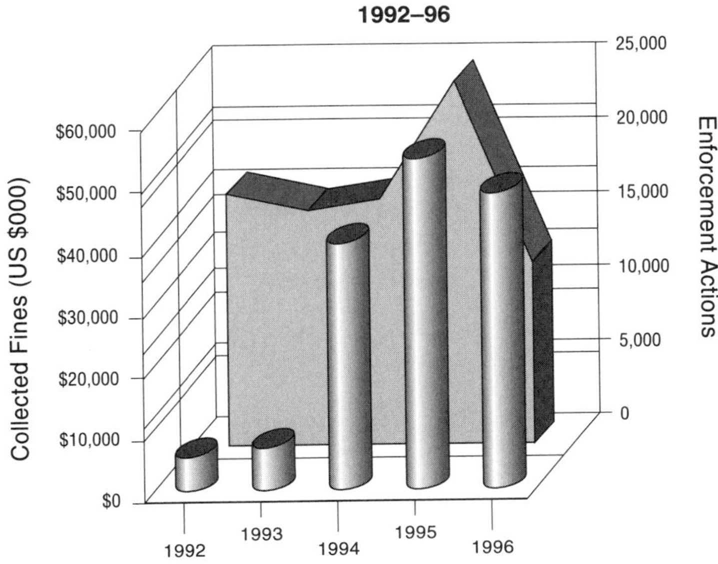

FIGURE 1.3 Enforcement of Natural Resource Protection Laws in Brazil: 1992–96 *Source:* IBAMA.

agency, competing for general revenues and personnel. Disputes with the newly created Ministry of the Environment regarding roles in policy formulation and execution only created further cause for dissension.

During the last few years, however, the fines, previously insignificant as an incentive to environmental performance, have increased in value, and their applications have increased in frequency. Table 1.1 and figure 1.3 show how infraction notices and fines have been meted out by IBAMA over the 1992–96 period on a sectoral basis. (There are as yet no unified registries of state-level enforcement activity; therefore, this listing considerably understates industrial pollutant control actions, which are primarily a state responsibility except in cases where pollution affects interstate water flows or air quality.) Despite the apparent increase in enforcement activity, only about half the infractions have resulted in fines actually being ordered by the courts. Fines still accounted for only about 16 percent of IBAMA's total operating expenses in 1994 (IBAMA/DIRCOF/DEFIS 1996).

The chronic shortage of resources has been in part overcome through external financing, although the national counterpart funds

Table 1.1 Sectoral Enforcement Actions by IBAMA: 1992–1996

	Year	*1992*	*1993*	*1994*	*1995*	*1996*
Forests	Infractions	14,700	8,424	12,219	20,315	8,239
	Fines	n.a.	n.a.	36,592,742	50,492,325	16,474,625
Wildlife	Infractions	392	3,815	2,449	1,138	941
	Fines	n.a.	n.a.	1,949,246	485,107	578,463
Fisheries	Infractions	2,057	1,599	1,664	2,285	2,304
	Fines	n.a.	n.a.	452,421	966,211	1,134,736
Pollutants	Infractions	60	2,127	321	830	672
	Fines	n.a.	n.a.	559,413	1,404,302	1,628,150
Totals						
	Infractions	17,209	15,965	16,653	24,568	12,156
	Fines	$5,221.3	$6,391.6	$39,553.8	$53,347.9	$47,559.1

From IBAMA/DIRCOF/DEFIS (1996 and prior years).
Values in current $1000. Values for 1996 have been annualized as an approximation.

required to match these have been severely restricted by fiscal policies. Royalties derived from natural resource extraction in Brazil are only partially applied toward abatement of environmental spillovers. In the same way, revenues derived from timber extraction taxes, from sewage system user fees, and from other charges are so minimal that they offer little incentive to restrain resource extraction or discharge. Where the real threat of enforcement is applied, the danger of corruption of enforcement officials represents an added impediment.

Despite the minimal effectiveness of economic instruments to date, evidence exists in Brazil and other Latin American nations to suggest that their use can be effective both in motivating technology adaptation to reduce environmental damages and in generating additional revenues to finance monitoring and enforcement (Serôa da Motta and Reis, 1994). A case in point is that of pollutant-intensity sewage charges on industrial users established in the 1980–82 period by the state water and sewer authority of São Paulo, SABESP. This system was devised as a means of raising revenues to subsidize households. When industries found they were being charged on the basis of the intensity of biological oxygen demand (BOD) or suspended solids discharges, they moved rapidly to adopt pollution-reducing technology to cut their charges. However, this resulted in SABESP no longer being able to generate the funds needed for its incremental operating expenses. The progressive charge system was hence abandoned and a block water and sewer charge reinstated, which removed the incentive to reduce pollution loads.

An important additional example of unanticipated responses associated with use of economic instruments in natural resource policy is detailed in the case study of charcoal production by Medeiros (chapter 4).This case refers to the recent adoption by the government of Minas Gerais state of a progressive tax on charcoal purchased by steel industries, to motivate a switch from natural to industrial forest plantations as a source of reducer fuel. A tax differential that quintupled this cost for charcoal derived from deforestation led many firms to switch to imported mineral coke (often more polluting than charcoal), to expand the radius of charcoal purchasing to neighboring states, or to threaten plant closure. This case plainly illustrates the difficulties in assuring the effectiveness of market mechanisms: Their repercussions may go well beyond the limits of the endangered systems that are the object of protection.

The control over industrial pollutants has generally been accomplished in Brazil through command and control measures such as standards and penalties. The effectiveness of these instruments has been diverse among Brazilian states. The most wealthy states in the south and southeast regions of Brazil's industrial heartland have benefited from both institutional arrangements and public mobilization in the environmental arena for the level of control they have attained. Given this differentiation and poor monitoring and enforcement capacity, government authorities have tended to resort to a more intensive environmental review process at the licencing phase for new or modified installations, rather than depending on ongoing monitoring of pollutants.

In general terms, the procedures used for environmental licencing of polluting industries follow the system developed in the late 1970s by FEEMA in Rio de Janeiro and, in 1981, at a national level (SLAP). Such licencing provides authorization for industries to install, expand, or initiate operations that may potentially cause environmental damages. The aim of the SLAP is to ensure compliance with publicly articulated environmental objectives, while at the same time motivating firms to accommodate social interests at least cost. However, over time, the procedures have been modified and relaxed, in response to the vagaries of public administration. Currently, the Ministry of the Environment seeks to revise and harmonize these procedures. Each state is therefore engaged in a diagnosis of their respective SLAPs, with a view toward revision that would reduce the inconsistencies and delays in licencing that have been serious stumbling blocks to rational application of environmental codes.

The SLAP operates through a three-tiered process that includes a period of initial project licencing during the planning phase (valid for two to three years), and proceeds through licencing for installation (three to six years) and operation (five to ten years). Although it was initially required only for industrial projects, after promulgation of the rules defining EIA/RIMA in 1986, environmental licencing became mandatory for a far broader range of activities, including residential subdivisions, tourism complexes, roads and railways, pipelines and sewage interceptors, electrical transmission lines, quarrying, and agricultural and timber extraction over 50 ha. The states have formed environmental commissions whose function it is to review norms and procedures adopted, review and approve licences, apply penalties, and oversee the performance of the environmental control agencies. In most cases, the commissions delegate to the respective control agencies the function of reviewing and issuing licences. Despite intentions to reduce the complexity of processing such licences, the capacity of environmental agency personnel to manage this process is severely limited in most states, and extreme delays are common.

Although the SLAP procedure specifically provides for application of penalties for noncompliance, in practice public agencies are not able to perform thorough monitoring. For example, in São Paulo, where environmental control measures have a long history, public complaints continue to guide the government agency responsible for environmental enforcement—CETESB—to those industries that may be violating standards. After reviewing compliance, the agency advises the firm if it will be necessary to bring its process technology into line, usually providing one month's notice for some indication that the necessary measures are being initiated. If standards are still in violation after three repeat monthly inspections, subject to progressively increasing fines, the firm may be shut down by the state environmental secretariat. Although the fines themselves are relatively small in comparison to the cost of adopting control technology, firms have often found that it is in their interest to avoid consumer dissatisfaction or legal battles in the public eye.

Another instance of accommodation to make environmental compliance more realistic within the Brazilian context is the adoption of "self-control" mechanisms. Through this program as it was adopted in Rio de Janeiro (PROCON), for example, firms themselves take on the responsibility for monitoring effluents and pollutant emissions, at a level of precision and regularity, and with chemical analysis undertaken by laboratories approved by the responsible public agency. Perhaps the most extreme case of self-regulation is that of Bahia, whose environmental secretariat has adopted an even more laissez-faire attitude, assuming that firms will find it in their interest to adopt environmental safeguards and report their progress toward management objectives periodically to the state government. This level of deregulation has been considered too lax in some quarters. Finally, environmental auditing procedures have been adopted in some states and would become compulsory should proposed federal legislation be approved. Such procedures would facilitate environmental monitoring and improvement within the context of the technological and market conditions facing each firm, but they represent an additional cost burden at the outset.

Even without major enforcement of state regulation, there has been a growing response by exporting industries to mounting pressures in international markets to adapt quality norms to include environmental factors. The adoption of quality and environmental management processes promoted by the Industrial Standards Organization (ISO) through the ISO 9000 and 14000 series, and the adoption of eco-labeling, sustainable source, and organic certification criteria have not gone unnoticed by Brazilian firms whose actions affect the environment. Such mechanisms, although voluntary, in practice affect prospects for sectoral competitiveness in an ever more interdependent market system. Thus a combination of regulatory and market instruments are in increasing evidence as components of a unified toolkit for environmental and natural resource management in Brazil.

METHODOLOGICAL ISSUES IN ECONOMIC VALUATION AND POLICY ASSESSMENT

To define optimal charges and taxes in market-based instruments, as well as to justify investments in environmental maintenance and recovery, it has become increasingly important to assess the social costs of resource depletion and loss of services. The recognition of these costs has recently led to the adoption of environmental valuation in the investment review procedures of the principal international development banks and of methods for inclusion of "satellite" environmental accounts in the framework of the unified System of National Accounts administered by the United Nations Statistical Office (Bartelmus 1996).

This section describes some of the theoretical and methodological issues pertinent to the economic valuation of natural resource and environmental quality, and the choice among regulatory policy instruments.

The theory of resource use founded in neoclassical economics considers that, if firms equate marginal returns to the marginal costs of resources employed in productive processes, they will be assured the maximum in profits with minimal utilization of the most scarce resources. Implicit in this equation is that the most scarce resources are those most valued. This allocative mechanism results in definition of the parameters of scarcity by the market price of factors and products, considered to reflect consumer preferences and production possibilities within resource constraints.

These concepts of allocation and revelation of preferences, fundamental to mainstream neoclassical economic theory, are not adequate for the valuation of goods whose price is not evidenced in markets, and whose users, including future generations, do not have the means to make their preferences felt in the marketplace. These problems lead to market failure, resulting in overutilization of natural resources and pressure on the capacity of natural ecosystems to absorb residuals. Based on this reasoning, environmental and natural resource economists have proposed that the market mechanism be adjusted so that such failures are internalized by economic actors. This analysis has led the State to apply policies that seek to adjust market parameters through instruments such as charges, technological restrictions, direct investments, and incentives to economic actors, so that they come to observe the carrying capacity of the physical environment. From this perspective, economic evaluations of environmental losses are undertaken so that public agencies may correct market signals to "get the price right" (Pearce and Turner 1989).

The initial assumptions in such research are that markets are reliable institutions, that they are already pervasive, and that solutions to the problems caused by their failure can be arrived at within them. The appropriate line of action is not always clear cut, however. Indefinite access rights to resources are a principal source of inefficient resource allocation and remain ubiquitous despite the rapid expansion of formal markets in developing economies. Nobel prizewinner Ronald Coase's influential market solution to external environmental costs (Coase 1960) suggested that negotiation between actors in a market context would be efficient and symmetrical for all actors regardless of the assignment of property rights. But recent work by Chichilnisky and Heal (in press) suggests that efficiency in dealing with certain types of environmental goods (those they define as "privately produced public goods") requires that equity principles be integrated with market allocation. Distributive justice is important to achieving allocative efficiency where environmental goods are concerned.

In contrast to a market-centered approach, many environmentalists and political ecologists treat the entire idea of monetization of natural services as inherently abhorrent (Martinez-Alier, in press). They argue that since it was the market that got us all into this mess in the first place, it would be suicide to let the market take care of the solution. Contending that humans have no predetermined rights to pollute, or to trade in services that the environment provides as a gift of nature, they propose that only physical limitations on human activity be applied on the basis of scientific knowledge and regulatory action.

There is thus a growing concern from some fronts that the corrective prognosis derived from neoclassical economics theory may represent a technologically optimistic approach contrary to the prudent skepticism needed to trace a sustainable future (Costanza 1991). The environmental problems that have arisen over the past few decades have taken on a global significance, and they are both cumulative and intertwined, having possibly catastrophic implications (although scientific certainty is lacking as to their causes and eventual solutions). As such, they are predicted to require collective action to be overcome, but they involve actors whose values are frequently in conflict. Following this line of reasoning, problems of this nature require a new approach to scientific reasoning that would be systemic, transdisciplinary, and "postnormal," requiring the direct involvement of affected actors to negotiate solutions to socioenvironmental conflicts (Funtowicz and Ravetz 1994).

As a result of cumulative transboundary effects associated with current environmental problems, these authors argue that a macro perspective is needed of sustainable development pathways, one that contrasts sharply with the micro, firm-oriented approach of traditional environmental economics. A macroeconomic theory of sustainability, according to Daly and Cobb (1989) and Daly (1992), should include three objectives simultaneously: *efficiency, equity,* and *scale.* Neoclassical economics has not devised analytical methods to incorporate objectives beyond that of efficiency, leaving the distribution of income as a fixed parameter (but see Chichilnisky and Heal, in press), and nature as only one among several sources of inputs whose marginal contributions should be equated for the efficient functioning of the market. Inequities of distribution in access to resources and in the direction of income derived from their utilization is customarily treated as exogenous to the analysis, with the argument that the weighting of concerns (differential utilities) of different income groups would distort the central objective of development, i.e., economic growth.

Daly (1992) proposes that the priorities be completely inverted so as to, first, ensure that a clear definition is made of the appropriate *scale* for growth in the physical occupation of space by the economy within biosphere limits. The economy, in this sense, is seen as an open subsystem of a finite and closed ecosystem. For scale to be sustainable, it must not diminish the capacity of the environment to support human-induced pressures. Expressed in more readily grasped neoclassical terms, the goods and services derived from nature, and their prospective availability in the future, are worth more at the margin than the benefits derived from a marginal growth in scale resulting in increased exploitation of those resources in the present.

Only once the appropriate scale is defined, and the rules are established to stay within its limits, can society define forms to equitably distribute rights to use and abuse nature. And only once there is an adequate definition of scale and norms of equity can the market be left to carry out its task to efficiently allocate goods and services in response to the laws of supply and demand.

These concepts that orient the transdisciplinary area of ecological economics do not in principle exclude analytical methods practiced by neoclassical economics. Rather, ecological economics relies on a variety of means to assess natural resource values, based not only on markets or "pseudo-markets" but also on the perceptions of users and interest groups of resource scarcity and uniqueness, and on ecological assessment of fragility and resilience. Also evident in this approach is the precedence of sociopolitical process over technocratic determination of society's future.

There are thus a number of factors that create distinctions between the various analytical approaches to natural resource and environmental valuation. Controversy exists with regard to whether compensatory investments against natural resource depletion should be made, and whether such investments should go to natural, man-made or human capital (the "strong" versus "weak sustainability" debate; see Jansson et al. 1994). This controversy pits mainstream economists who treat infinite natural resource substitution by man-made capital or technical innovation as inevitable and desirable, against the ecological economics precept that would adopt clear-cut limits to the scale of the human economy within ecosystem constraints. As a result, controversy also persists among analysts regarding which monetary values are the most appropriate to adjust the Gross National Product so as to reflect the costs of resource depletion and contamination. This unfortunate dispute has confounded the interpretation of such values as a framework to guide national development policy.

Although these debates have impaired the successful dissemination of national environmental accounts, a number of revealing initiatives have been pursued, notably in Latin America.[2] As yet, the Brazilian government has made little progress toward the integration of environmental accounts, despite procedures and estimates advanced by analysts at the Ministry of Planning (Serôa da Motta 1995; May in chapter 2 of this volume). Methodological controversy has discouraged the implementation of environmental accounts initiatives. In response, some have now turned to the promotion of "indicators of sustainability" as an alternative, based on physical progress toward socially consensual objectives for natural resource status (Winograd 1995), rather than adhere to monetary accounts. Such a response does not vitiate the need for refinement in monetary valuation techniques, that offer important contributions to the assessment of alternative investments, the levying of taxes and charges against resource depletion, and the estimation of damages in liability suits. It is with these objectives in mind that the contributions in this book are directed.

A CRITICAL OVERVIEW OF METHODS, AND CONCLUSIONS FROM THE CASE STUDIES

The studies gathered in this volume represent efforts to value previously unpriced resource goods and services in Brazil from the perspective of both today's and tomorrow's users, as a contribution to policy formulation. The focus of the studies is primarily on choice among competing natural resource uses, although several studies also address policy mechanisms and socioenvironmental benefits associated with promotion of cost internalization and compensatory investments in urban-industrial contexts.

The methods applied in the assessment of natural resource values range from those derived through direct use of wood and nontimber forest and agropastoral products in market transactions (see chapters 2 and 3) to the contingent valuation of indirect resource benefits (primarily sport fisheries) derived from maintenance of intact wetland ecosystems (see chapters 5 and 6). A more hybrid approach, that estimates a "total economic value'" comparing natural versus plantation forest extraction for charcoal manufacture, is provided in chapter 4.

The "user cost" approach adopted by May (chapter 2) suggests that a portion of the returns obtained from resource use in the present be invested in capital whose constant flow of benefits can be captured once the original resource is depleted—an application of the Hicksean concept of sustainable income (El Serafy 1988). In the case of Brazil presented here, this approach leads to values for Amazon forest resource depletion that, although of significant global concern, are surprisingly low in an accounting context based on sustainable timber and nontimber production foregone, because of their apparent abundance (lengthy depletion horizons). On the other hand, the application of a "net price" approach to the forest accounts would suggest that Brazil's agricultural frontier expansion over the past decades led to costs greater than the value added by the entire sector, a result that would urge total revamping of national agricultural policy to avoid any additional deforestation. Although this latter result would please some environmentalists, it is clearly unacceptable in the light of the concerns of the landless peoples' movement.

The user cost approach employed in natural resource accounts estimation offers a temporal perspective on the sustainable income that may be lost through rapid depletion, but it also more transparently reflects the definition of ecological limits to expansion in the geographic scope of the human economy. Reduction in the exploitable resource stocks through zoning, for example, would dramatically enhance the valuation of Amazon forests because of the ensuing scarcity attributed to timber and potential agricultural land uses (May 1996).

The analysis of natural resource benefits commonly assesses the temporal pattern of realization of those benefits, and the discounting of such flows to the present. This approach has been criticized as inequitably favoring the demands of the present and bankrupting future generations. Thus choice among resource exploitation options favors those flows that offer greater up-front net returns to the investor or society, while those choices which would guarantee long-term sustainable flows would be shelved. It may therefore come as a surprise to readers that the comparative analysis of land uses in the eastern Amazon presented by Almeida and Uhl in chapter 3 demonstrates that, even when market discount rates are applied to the resource user's future earnings, long-term, more sustainable approaches (managed timber extraction, pasture renovation, and agroforestry practices) are financially superior to "mining" timber and soil. The problem, of course, is that public policies distort the benefits perceived by resource users, leading them to take the more immediate but less sustainable path.

A more comprehensive social benefit-cost analysis, which would remove the effect on present earnings of distortionary tax and credit incentives to the cattle industry, more adequately prices timber at its stumpage value rather than as a nearly free good. As it measures rather than ignores loss of environmental services and of biodiversity resulting from deforestation, it would demonstrate even more fully the need for a broad shift in sectoral policy to redress such distortions.

In this sense, the study by Medeiros (chapter 4) seeks to assess monetary values attributed to a more complex array of environmental impacts, although the temporal depletion horizon is not incorporated as a factor in valuation. His study compares the use of alternative wood sources as raw materials for charcoal manufacture in the Minas Gerais steel industry. The state environmental policy to limit deforestation in the sparsely vegetated and fragile savanna plateau led to imposition of a highly differentiated sales tax on charcoal from native timber as opposed to that derived from plantations, chiefly eucalyptus, that had previously been left standing in favor of the cheaper native stands. Both sources, however, have environmental implications that were not adequately valued in considering the impacts of the tax policy. Loss of native stands leads to soil degradation, net CO_2 emissions, and loss of biodiversity, effecting displacement of rural workers who formerly relied on forest product extraction for part of their livelihood. But eucalyptus plantations also erode soils rapidly during establishment, impoverish species diversity, and may substantially compromise subsurface water supplies. The exercise in monetary valuation actually showed higher cost to society on the basis of the charcoal derived from eucalyptus plantations—suggesting this charcoal should perhaps be taxed more rather than less—although the cost per ha of native forest lost was considerably higher than that associated with an equivalent area put into eucalyptus. These results imply that more needs to be done to stimulate appropriate rural land use than merely to differentially tax alternative sources of fuelwood used in charcoal manufacture.

As mentioned previously, the use of the "sustainable source" criterion to differentiate pricing of charcoal from native versus plantation origin had unanticipated side-effects, both economic and environmental. The displacement of charcoal production to neighboring states that had not yet imposed equivalent taxation engendered higher transport costs and petroleum fuel emissions. Steel foundries discovered that cheaper fuel could be had by importing mineral coke from China, whose high sulfur content and particulate emissions exacerbated air quality and provoked respiratory illness. However, some of the larger integrated steel manufacturers found it advantageous to expand their eucalyptus plantings, neighboring states are actively pursuing equivalent tax legislation, and the global coke market is no longer as obvious an alternative to domestic charcoal.

The following two studies address the unpriced values attributable to natural wetland environments, as assessed by nonconsumptive users of mangrove and freshwater wetland ecosystems in Brazil's coastal estuaries (chapter 5) and Pantanal (chapter 6), respectively. Both studies make use of the Contingent Valuation Method (CVM). This approach is one of a few widely applied tools that, in the absence of markets, make use of so-called surrogate markets to elicit consumers' willingness-to-pay to maintain flows of environmental goods and services. In practice, actual or potential beneficiaries of environmental goods are asked to define how much they would pay in monetary terms to ensure maintenance of such goods and services; alternatively, they may be asked how much they would accept to go without them (in theory, the two approaches are symmetrical).

A number of difficulties clearly emerge in the application of contingent values (Hannemann 1996). One is that some consumers may tend to overstate their possible true demand for environmental quality, until it comes time to actually pay rather than simply express willingness to do so. Others may be willing to pay a stated amount for a given good but, because of budget constraints, would limit their actual allocation of income were they then to be asked to pay for a wider array of environmental benefits. Similarly, people who live in poverty cannot be expected to do without immediate income to ensure that others are assured of a quality environment (although analysts are often surprised by the proportion of even meager incomes that poor people express themselves willing to set aside to protect natural values). And finally, although theoretically symmetrical, people are generally much more apt to accept compensation for loss than to pay for receiving environmental services.

Despite these limitations, and the tremendous statistical problems inherent in estimation of hypothetical demand, CVM approaches are today among the most often used techniques to assess values for unpriced environmental services. The principal multilateral development banks regularly have recourse to this approach when evaluating projects whose benefits flows are primarily obtained from investments in environmental quality improvement. The infamous assessment of penalties to compensate damages from the Exxon Valdez disaster in Prudhoe Bay off the Alaska coast was founded on contingent valuation of societal losses as well as of ecosystem recuperation costs. The immensity of the judicial penalties awarded in this case led the U.S. National Oceanic and Atmospheric Administration to propose guidelines toward such damage assessment, adopting as a rule of thumb a maximum of 50 percent of the contingent-valued damages.

Critics of the approach stress that the "chrematistic" assessment of monetary values to goods and services that may not in fact have use value, but rather emotional or symbolic importance, unjustly homogenizes and impoverishes their cultural designation. Furthermore, it would be inequitable to those yet unborn to attribute values assessed by current users to the benefits of unidentified ecosystem services such as biodiversity, which may actually generate use values to society only in the very long term (Martinez-Alier, in press).

The final two chapters of this volume deal chiefly with urban-industrial questions associated with water pollution and with human settlements at risk. Serôa da Motta and collaborators (chapter 7) offer two case studies of valuation appropriate to the determination of investment requirements and internalization of pollution damages, while May (chapter 8) assesses environmental benefits obtained from public investment in community reforestation of degraded slopes that endanger urban squatter settlements.

The assessment of health damages arising from exposure to pollution represents another widely used approach to the valuation of environmental damages. The case of waterborne disease estimated by Serôa da Motta and Rezende (chapter 7) reflects an evolution in Brazilian thinking regarding the appropriate measures to use in valuing lives saved by investment in basic sanitation services. A prior study by analysts at the same institution, for example, had recourse to so-called human capital valuation, in which environmental costs are estimated on the basis of societal labor output foregone (valued in terms of expected future earnings) of those who perish as a result of illness spread through inadequate sewerage or air quality control (Serôa da Motta and Mendes 1995).

Human capital values have become extremely controversial, as was apparent from debates ensuing from divulgation on the eve of UNCED of an infamous memorandum by then Chief World Bank Economist Lawrence Summers (now Secretary of the U.S. Treasury). In that memorandum, Summers argued for the "impeccable" economic reasoning behind dumping of toxic waste in developing nations due to the lower cost attributable to mortality among their peoples, vis à vis those of industrialized nations. Needless to say, the memorandum was widely condemned, even by the conservative press (such as *The Economist,* which gave the issue considerable airing at the time).

In the present study, Serôa da Motta and Rezende derive statistical relationships associating mortality with the lack of basic sewerage and adequately treated water supplies, common in many parts of Brazil. Monetary values are then derived based on the imputed number of lives *saved* for a given marginal investment in sanitation services, a value that can be of great significance for investment choices.

Serôa da Motta and Mendes, in the second case in chapter 7, assess the marginal costs attributable to water pollution control investments by industrial facilities as a means to simulate the effectiveness of effluent taxation on technology adoption. This analysis relies on the pollutant taxation policy advanced by Cambridge economist Pigou in the 1930s. Simply put, the pigouvian tax would equate the social costs of pollution with the firm's marginal revenues derived from additional emissions, finding a "socially optimal pollution" level. The difference between social and private costs at this point would indicate the amount of taxes that should be levied against firms' emissions so as to induce them toward more acceptable pollution levels. Firms able to reduce emissions at a cost per unit emitted lower than the tax rate would do so, while those with more costly pollution control would be better off paying the tax. Serôa da Motta and Mendes' simulation shows that the social costs of such market-based measures would be considerably lower than those associated with a uniform emission ceiling across firms.

The relative efficiency of economic instruments over command-and-control measures is fine in theory, but in practice there has been substantial opposition to adoption of such measures, since in some cases they are perceived as being tantamount to offering a licence to pollute. As described in the cases of effluent charges in São Paulo and charcoal taxes in Minas Gerais, the only experiences of this kind in Brazil have had the unwanted side effects of crimping regulatory agency revenues or inadvertently stimulating other environmental problems. This suggests that economic measures for environmental management be carefully tested at a pilot level before being broadly implemented.

The final case in the collection describes the experience of Rio de Janeiro in combating socioenvironmental conflicts that ensue from the occupation of steep hillsides by squatter communities (*favelas*), and clandestine granite quarries. In this study, May presents a monetary valuation of environmental benefits derived from investments in reforestation of unstable slopes by a local government agency in collaboration with a residents' association that mobilized sweat equity. The case details how political mobilization of *favela* dwellers led progressive administrations to adopt them as neighborhoods rather than illegal settlements, and to change tactics from violent eviction and relocation to the provision of urban services and environmental restoration.

In economic assessment of the socioenvironmental benefits from compensatory investment in slope stabilization, May proceeds to compare the costs before and after reforestation to residents and neighboring communities associated with risks of flooding, water shortage, and landslides. Reforestation also improved property value and increased recreational options. The analysis established the superiority of such benefits over the costs involved in tree planting and minor drainage works, providing ammunition for the public agencies engaged in this effort to seek budget allocations that enabled expansion of the program. Furthermore, the evaluation stressed community mobilization as critical to success, since the program adopted environmental education and prioritized selection of unemployed workers as employees, guaranteeing adoption of the reforested areas and works as community property.

On the other hand, equity analysis of these public expenditures suggests that the greatest beneficiaries of such projects are downstream communities, no longer exposed to sedimentation and consequent flooding, whose properties increased in value with the views of verdant slopes that replaced the scars of slums and quarries. The squatters themselves pronounced themselves far more concerned with provision of basic sanitation services and security against the violence of drug trafficking that permeate their day-to-day lives, rather than the reduced risk of eventual landslide. In this respect, the resolution of socioenvironmental conflict must be more informed by the priorities of the community of interest, than by the evaluation of purported monetary benefits to the broader society.

NOTES

1. Recent modifications in the rural land tax legislation provide for the exemption of forest lands declared as protected reserves. IBAMA has also promoted the establishment of so-called private nature protection reserves, which provide for permanent exemption of lands ceded in perpetuity for conservation.
2. A case in point is that of Chile, where criticism of the pulp and paper industry that aggressively substituted native forests with plantations raised international attention over the potential political sensitivity of environmental accounts (unfortunately resulting in the dismissal and blacklisting of Central Bank analyst Marcel Claude, who had been responsible for those accounts).

2

Measuring Sustainability: Forest Values and Agropastoral Expansion in Brazil

PETER H. MAY

SUSTAINABLE DEVELOPMENT INDICATORS

To measure progress toward sustainable development, one must first define the goal that is being pursued. While private firms perceive economic progress as being measured by profits and financial rates of return, governments use employment and growth in GNP as key measures of economic health. For neoclassical economists, these indicators and their maximization objectives are mutually consistent: Individual profit maximization favors national income growth and full employment of productive resources, including labor.

On the other hand, the financial rate of return desired by private enterprise may not lead to sustainable rates of growth, because of market failures leading to exhaustion of the resource base and clogging of the sink capacities of the natural environment. Traditional measures of economic product treat resource extraction as income without compensating for the drawdown on natural capital, and expenditures on pollution control and waste cleanup are likewise bundled together with gross product.

Efforts were begun in the late 1980s to devise methods for adjusting national accounts to reflect resource exhaustion and environmental services (Repetto et al. 1989; Solárzano et al. 1992; Bartelmus et al. 1992). Although all such accounts must rely on adequate indicators of change in resource conditions, there are a number of disparate approaches to environmental accounting. The differences between these methods stem from divergent perspectives regarding the relative validity of substituting natural for man-made capital in the search for sustainable development. That their results sometimes diverge by enormous magnitudes suggests that debate over differing indicators to measure sustainable development reflects substantial difference of vision regarding the definition of sustainability.

A "weak sustainability" definition assumes that a range of possibilities exist to substitute natural for manmade capital. Sustainable income in this view represents the amount that can be consumed so that "at the end of the day" one is as well off as at the beginning, a concept derived from John Hicks' (1946) definition of *income*. Referring to exhaustible resources, the "Hartwick Rule" provides that one must reinvest rents (called "user costs") from exploitation of natural resources so as to achieve constant real consumption over time, if such extraction is not to be construed as a loss in intertemporal welfare (Hartwick 1977). Where one reinvests the proceeds from resource exploitation is determined by the neoclassical precept that one puts one's money where it earns the best return. There is no constraint in this approach that one must dedicate part of this rent to cleaning up the mess that was caused by the exploitation in question, and there is no particular incentive in the structure of user costs that would motivate a change in the rate or form of exploitation.

Although various gradations may be foreseen,[1] a simplified version of the "strong sustainability" argument suggests that substitution prospects are not at all infinite, and that development is sustainable when at least some ecosystems remain intact. These include those necessary for the provision of "life support" functions, such as maintenance of carbon balance, hydrologic cycles, and nutrient flux (Pearce and Atkinson 1992). This perspective does not prohibit resource exploitation but suggests instead that there should be investment of resource rents in *natural* capital, not just any capital that would provide income in the future, so that the net change in the stock of these resources would be greater than or equal to zero. One measure of sustainable income derived from this approach is the "net price" or depreciation approach followed by Repetto et al. (1989), in which the full value of resource rents (market prices less transport and extraction cost) is deducted from net product to provide a measure of resource depreciation.

To consider the implications of these differing perspectives on national environmental accounts, analysts at Brazil's Institute for Applied Economic Research (IPEA), prepared a series of estimates of such accounts applied to Brazilian mineral and forest resources, and to environmental services derived from water and atmospheric sinks for urban residuals (May 1993, 1994; Serôa da Motta and Young 1991; Serôa da Motta et al. 1992; Serôa da Motta and May 1992, 1996). This chapter describes the results of those studies that focused on deforestation at Brazil's agropastoral frontier. The analytical framework for user cost and net price values derived in this study is provided in appendix 2.1.

Despite the rural resource emphasis of this discussion, readers should bear in mind that Brazil is a country that is overwhelmingly urban (75 percent of the 1991 population) and industrialized (nearly two-thirds of exports are manufactures). Thus recent attention to deforestation at best describes a small segment of the environmental costs of Brazil's development model. Indeed, for the majority of Brazilians, the environment is a question of poverty. Over 40 percent of the national population now earns less than what is necessary to fill a minimal market basket, most homes lack sewage collection or treatment, and infant mortality at 57 per 1,000 is still unacceptably high. Resource exploitation has not for the most part benefited the poor, since the majority of agropastoral expansion has been oriented toward increasing export production rather than toward reducing urban consumers' food bill.

BRAZILIAN FRONTIER EXPANSION AND LAND DEGRADATION

Brazil's national self-image is nurtured in its pride for expanding frontiers, wide horizons, and unlimited natural resources. Because of its continental scale and abundant mineral, water, land, and human resources, Brazil was able for many years to pursue a fairly autarchic economic development model. Wealth transferred from extensive export-oriented agriculture financed the nucleus of a substantial industrial sector based on subsidized hydroelectric power, nationalized steel and petroleum production, and a cheap, well-disciplined laborforce.

By 1980, Brazil's population was already predominantly urbanized and had a large and growing auto industry geared to the domestic market. To fuel the fleet, it was pumping large volumes of pure ethanol from expanded sugarcane plantations subsidized by taxes on gasoline, the latter distilled mostly from imported petroleum. To release the pressure valve of regional inequalities, and to offer yet another symbol of national manifest destiny, the government invested in the construction of a new planned capital at Brasília in the heart of the central savanna, and later cut new highways and railroads for agricultural settlements and mines deep into the dense Amazon forest.

Unlimited horizons and inadequate heed to environmental costs has resulted in a history of failure to observe biophysical limits in the headlong rush toward economic development. Pharaonic projects have placed Brazil among the ranks of the most severely indebted nations while at the same time nourishing environmental controversy. Among recent policy disasters were those related to the expansion of Amazonian hydroelectric generation, whose immense reservoirs (needed to compensate for the flat terrain) flooded out indigenous groups and diverse tropical biomes. The POLONOROESTE land settlement scheme in Rondônia and neighboring Mato Grosso also garnered international acrimony as its herringbone-aligned farm-feeder roads were shown to be stimulating deforestation and soil degradation at a rapid pace, while placer miners and lumber mills encroached upon indigenous and biological reserves. Generous subsidies and tax exemptions for beef cattle ranching helped to provoke further destruction and were shown to be justifiable neither economically nor environmentally (Hecht 1985; Browder 1988).

From the time of colonization up to 1982, human occupation ("anthropic activity"), principally for agriculture and livestock rearing, had succeeded in modifying 22.5 percent of Brazil's national territory (Veloso and Gões Filho 1982). Less than ten years later, in 1991, anthropic areas had grown by half, constituting at least 32.1 percent of that territory (IBAMA 1991). This acceleration in territorial occupation occurred in an uneven manner in different regions, as a result of variations in ease of access, regional development policies, political patronage, and the characteristics of exploitable natural resources.

Brazil possesses a wide range of complex ecosystems, among which are one-quarter of the world's remaining tropical forests, which makes the nation an important storehouse of global biodiversity. Although savanna (*cerrado*) ecosystems may contain a greater breadth of species, the sheer exuberance and immensity of the Brazilian Amazon result in a greater emphasis in international environmental activism and policy discussion being placed on that region. Nevertheless, as figure 2.1 shows, rainforest vegetation types (termed ombrophylous) are found in both the Amazon and the Atlantic Forest biomes. The latter is by far the most threatened because of human territorial expansion; thus the discussion, in Brazil at least, tends to give more than equal time to this region, which stretches from the temperate

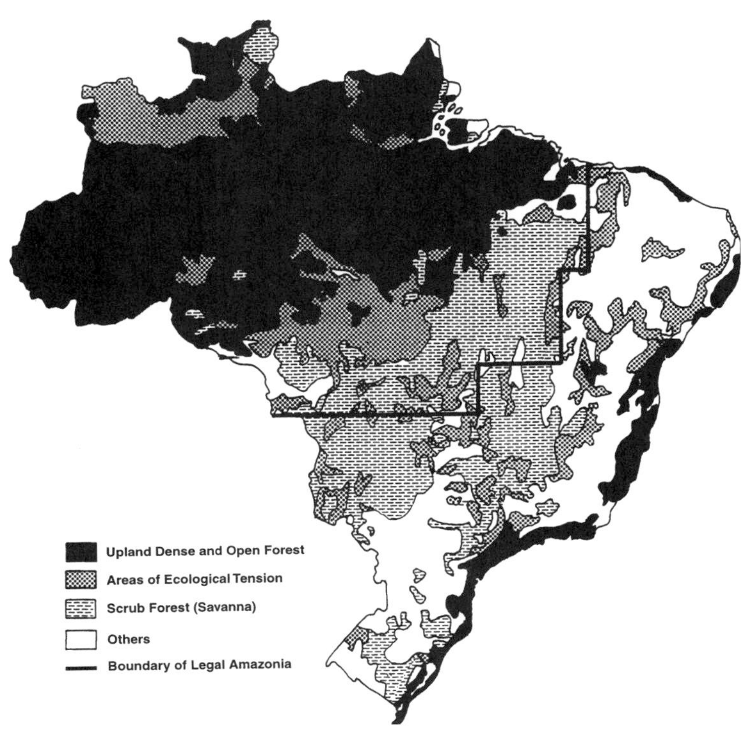

FIGURE 2.1 Principal Vegetation Types in Brazil ***Note:*** The Legal Amazon region lies in the area within the heavy black border.

coastal zone of Rio Grande do Sul to the equatorial beaches of Rio Grande do Norte. Altogether, the ombrophylous vegetation in the Amazon and Atlantic Forests is estimated to have originally covered an area of 3.4 million km^2 (IBGE 1982). In general, these forests are interspersed with other vegetation types such as cerrado and transitional ecosystems (*tensão ecológica*), so that strict definitions of vegetation stocks and boundaries are quite difficult.

Since colonization in the early 1500s, Brazil's occupation has been accomplished through removal of native vegetation and its nearly complete and permanent replacement by agropastoral land uses and, in far more limited area, by plantation forests of *Eucalyptus* and *Pinus.* These two species, grown in simplified monocultures, now cover an estimated total of 45,000 km^2, nearly all of which lies in what was once Atlantic Forest. Although Brazil's forest plantations stand

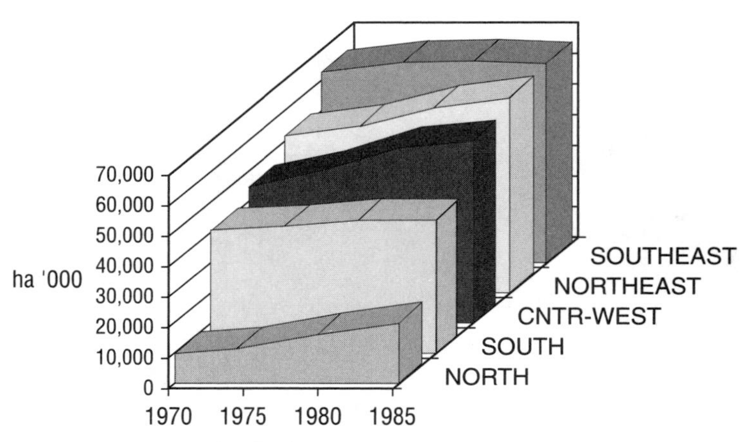

FIGURE 2.2 Expansion in Agropastoral Production Area by Region: 1970–85 (1,000 ha) ***Note:*** The figures refer to the total regional land area in agricultural establishments, including cropland, range and planted pastures, and industrial tree crops, but excluding areas under natural forest within those establishments. *Source:* IBGE *Censo Agropecuario,* various years.

among the world's largest reforestation efforts, they are dwarfed by the territorial magnitude of deforested area, and to them is attributed the deterioration of ecological services derived only from native forests. This suggests that it is not an easy matter to substitute natural forest capital, even with other "natural" capital investment.

Recent satellite imagery analysis shows that the species-rich Atlantic Forest that occupied much of the South-East region at the time of conquest has been reduced to less than 8.5 percent of its original coverage (SOS Mata Atlântica 1992). Most of this depletion occurred during the period between 1960 and 1980, a period known as the heyday of "conservative modernization" in Brazilian agriculture. Over these two decades, generous industrial and export crop subsidies stimulated rapid expansion of soybeans, coffee, silviculture, and cattle ranching.

In the ensuing decade, 1980 to 1990, deforestation in the Atlantic Forest slowed. The debt crisis beginning in 1981 diminished public support to agriculture except for the significant effort to expand biomass fuel production from the late 1970s, which also occupied remnants of the Atlantic Forest. Depletion rates in the 1980s averaged 1,000 km^2 per year in the entire region, down from over 7,000 km^2 in the preced-

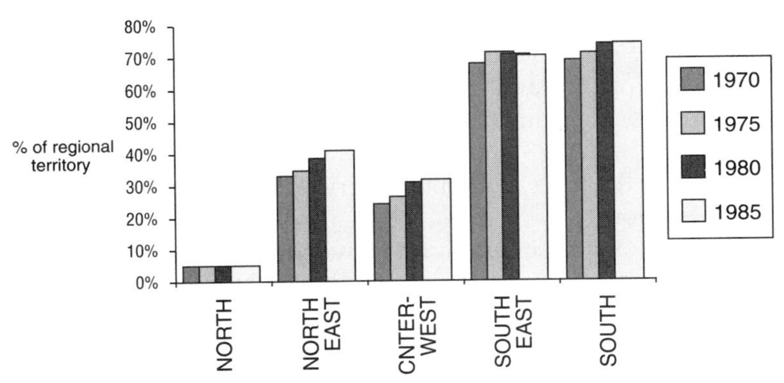

FIGURE 2.3 Expansion in Share of Territory Occupied by Agropastoral Production in Brazil, by Region: 1970–85

ing decade. Deforestation differed widely among states, with the most rapid depletion in absolute terms occurring in Paraná, Santa Catarina in the South, and Bahia in the Northeast. As early as 1970, in both the South and South-East macroregions,[2] agropastoral activities occupied 70 percent of the physical territory (see figure 2.3).

Agropastoral expansion in the Amazon (North region) has been the least significant in both absolute (figure 2.2) and relative area (figure 2.3), when compared with other regions in Brazil, but it is increasing more rapidly than that in long-settled regions. Devastation in the Amazon had, by 1991, reached a total of 476,000 km^2, or approximately half the total area deforested in the Atlantic Forest since the turn of the century. An additional 5,000 km^2 of original forestland in Amazonia is now under water—flooded by hydroelectric dams. A good part of this clearing occurred in response to subsidies for cattle ranching and public sponsorship of colonization projects and infrastructure development, primarily in western Maranhão, Pará, Mato Grosso, and Rondônia.

The rate of deforestation was considerably different in other regions. Upon reexamination of figure 2.2, it is evident that unforested agropastoral area has been increasing in the Center-West, the Northeast, and the North, while it has stabilized in the South-East and South regions. The rate of growth was particularly accentuated during the 1975–80 period, when debt-financed economic growth was in its heyday. At the outset of the 1980s, economic crisis was reflected in reduced credit to agriculture, causing a decline in the rate of frontier expansion. Nevertheless, in the Amazon and some parts of the Northeast, the prior pattern of expansion had taken on a life of its own, driven along by land speculation, gold fever, and the drug trade. These forces were dramatically portrayed in the burning that took place in 1987 and 1988 and was captured by satellite images widely distributed to the press, ushering in serious global concern for tropical deforestation.

Despite a policy response to restrain deforestation and burning by the Brazilian government through the "Nossa Natureza" program established in 1989 by the outgoing Sarney administration, the economic crisis of the late 1980s was evidently far more effective than the threat of fines in slowing agropastoral expansion at the frontier. When the crisis relented after the successful stabilization of 1994, the response was staggering. Data released in 1996 and early 1998 by the National Space Research Institute (INPE) showed that the rate of Amazon deforestation significantly increased in the 1991–95 period, after a slight decline from 1989 to 1991, the result, it was supposed, of increased monitoring and fines. Although 1996 brought a modest decline from this increased rate of deforestation, the rate reported for that year by INPE was the same the agency had discerned in the "burning season" of the late 1980s. These data represented a major blow to government in Brazil and led to a series of forest- and land-use policy shifts, as described in chapter 1. Deforestation is hence a persistent and growing phenomenon that merits continuing attention among a host of related and unrelated environmental issues.

ESTIMATING FOREST RESOURCE VALUES

By removing forests to make way for agriculture, livestock, and tree plantations, Brazil generates current receipts in substitution for the income that could have been generated had the natural forest been utilized for sustained timber and nontimber production. The returns from forest management and the indirect benefits that could have been obtained had forest cover been retained are ignored by the traditional System of National Accounts (SNA). This study concentrates on the benefits derived from on-site forest products. Off-site benefits such as watershed and microclimate stabilization, and global concerns such as carbon sequestration and biodiversity represent further challenges for resource valuation.

One of the reasons that forest losses are not adequately valued is the lack of information regarding the characteristics of forest resources converted for agropastoral purposes. This study sets out to characterize these resources in broad terms and to estimate what their productive value might have been had they been conserved.

In Brazil, forest inventories were accomplished in considerable detail during the RADAM program in the 1970s (Brazil 1973–1982; IBDF 1983), providing point estimates and regional extrapolations on species composition, range, and market potential. A vegetation map derived from these studies provided a basis for characterizing natural resources according to predominant forest ecosystems (Veloso and Gões Filho 1982). The principal mapping units (see figure 2.1) have broadly similar characteristics in terms of floristic composition, standing wood volume, and vegetation growth rates, because of similar climatic and soil associations. In this study, Brazilian states were classified as to vegetation type by superimposing political boundaries on this map and selecting the vegetation type that represents more than 50 percent of each state's surface territory.

Besides the fundamental description of general vegetation types, the study made no effort to differentiate timber stocks by species composition. In contrast, the inclusion of nontimber products involved the selection of six species whose output, valued at $687 million in 1989 (IBGE 1990), was responsible for nearly 50 percent of total nontimber extractive product value in that year.[3] These species are abundant in areas liable to be converted for agriculture. Species that might have important economic potential in the long run, such as medicinal resources, were not included, masking sources of genetic diversity whose value is only beginning to be recognized. Those species selected for analysis include native rubber (*Hevea brasiliensis*), Brazil nut (*Bertholletia excelsa*), babaçu palm (*Orbignya phalerata*), açaí palm (*Euterpe oleracea*), palmito (*Euterpe edulis*), and carnaúba palm (*Copernicia prunifera*), which are sources of gum, food, fiber, oils, and wax. Some, such as rubber, açaí palms, and Brazil nut, are found principally in the Amazon, while carnaúba grows extensively in the semi-arid Northeast, and palmito once occupied a wide coastal strip in southern Brazil.

The sustainable annual production of timber and nontimber products was obtained through consultation with forestry specialists and a detailed literature review. The study assumed that a sustainable management system would obtain, as the average annual permissible cut or harvest, a volume equivalent to the average annual natural increment in merchantable wood volume, or the natural yield of other marketable products. This volume is defined as that necessary to main-

TABLE 2.1a PRODUCTIVITY AND STOCK OF TIMBER RESOURCES

Predominant Forest Type	*Annual Wood Increment (m^3/ha)*	*Total Wood Volume (m^3/ha)*
Dense tropical forest	0.60	90.0
Open tropical forest	0.48	70.0
Atlantic forest	0.64	67.0
Caatinga (thorn forest)	0.20	37.5
Cerrado (savanna)	0.20	45.5
Semi-deciduous	0.50	67.0

From FAO (1985); Veloso and Gões Filho (1982); R. Jesus de Moraes and E. F. Durso (pers. comm.).

TABLE 2.1b PRODUCTIVITY OF NONTIMBER FOREST RESOURCES

Resource	*Productivity (kg/tree/yr)*	*Density (trees/ha)*	*Productivity (kg/ha/yr)*
Hevea brasiliensis (rubber)	1.06	2	2.12
Bertholletia excelsa (nuts)	100	1	100
Euterpe oleracea (fruit)	4.4	267	1200
E. oleracea (palm heart)	n.a.	267	600
E. edulis (palm heart)	n.a.	n.a	600
Orbignya phalerata (fruit)	30.4	53	1600
Copernicia prunifera (wax)	0.1	800	80

From Bovi and Castro (1993); Clement (1993); EMBRAPA (1976); Homma (1989); Jardim and Anderson (undated); Johnson and Nair (1985); May et al. (1985); MIC/STI (1982); Mori and Prance (1990); Renesto and Vieira (1977).

tain forest volume in equilibrium under climax conditions. Sustained forest management would imply removal of only this increment. This is a conservative estimate, since managed forests can be assumed to produce volumes well in excess of the natural increment. The annual increments and productivity values used in these estimates are provided in table 2.1.

Data on original vegetation coverage obtained from RADAM mapping were used to estimate forest stocks at the moment of resource inventory. Together with information on land-use change, these would enable calculation of annual depletion rates and hence the point of resource exhaustion. Despite the conceptual renewability of forest resources, this study assumed that conversion to agropastoral land use represents a permanent substitution—a process common in Brazil—thus withdrawing resources from stocks forever.

To identify changes in forest resources, information regarding land use within vegetation zones is necessary for more than one point in time. Unfortunately, there is little information of this type available from forest inventories for most Brazilian states. For this reason, this study took recourse to the Agricultural Census data, obtained at five-year intervals from 1970 to 1985, to calculate the approximate rate of alteration in forest cover within states.

The estimates of deforested area within agricultural establishments generated by this procedure enable approximation, both forward and backward from the inventory data points available, of annual loss in forest cover related to agricultural expansion and, assuming continuity of such losses at the same rate, the number of years (*n*) to resource exhaustion.

Finally, to complete the valuation of sustainable forest produce foregone, annual prices of timber and nontimber products were estimated. The rent value of timber resources at the point of extraction was approximated by deducting the average costs of extraction and transport from international market prices per unit output to obtain a stumpage (standing timber) value.

Based on the Brazilian input-output tables for 1975 (IBGE 1979), average stumpage payment to timber resource owners was found to be on the order of 75 percent of gross FOB sales value for wood products,[4] whereas the farmgate prices for nontimber products were estimated at 50 percent of their reported domestic value, because of the complex intermediation between producers and markets common for these products in Brazil. These proportions were applied uniformly over the fifteen years under analysis.

To summarize, the annual estimate of sustainable production foregone was derived for each state using the following formula

$$Y_{jt} = [(K_{jt+1} - K_{jt}) \cdot g_j] \cdot p_{jt},$$

where

Y_{jt} = sustainable forest production from resource *j* in period *t*
K_{jt} = forest resource stock in period *t*
g_j = productivity of resource *j*
p_{jt} = average price of forest product derived from resource *j* in period *t*.

The annualized total of sustainable wood and nontimber production foregone, derived using this formula, is presented in table 2.2 as an average for census years. These values are considerable, ranging from \$162 million to \$365 million for wood alone. Nontimber products

TABLE 2.2a SUSTAINABLE WOOD PRODUCTION FOREGONE, BRAZILIAN MACROREGIONS

Macroregion	*1971–1975*	*1976–1980*	*1981–1985*
North (Amazon)	$63,103	$113,852	$73,800
Northeast	$50,737	$74,512	$32,654
Center-West	$73,142	$116,961	$38,197
Southeast	$15,704	$17,493	$9,517
South	$36,952	$42,156	$7,721
Brazil	$239,637	$364,973	$161,890

From May (1994).
Amounts reported are the annual average in 1980 US$1000.

TABLE 2.2b SUSTAINABLE NONTIMBER PRODUCTION FOREGONE, PRINCIPAL SPECIES

Forest Resources	*1971–1975*	*1976–1980*	*1981–1985*
Açaí	n.a.	$2,051	$2,807
Babaçu	$47,678	$32,860	$25,078
Carnaúba	$349	$2,346	$943
Brazil nut	$2,174	$2,810	$2,299
Rubber	$122	$132	$31
Palmito	$1,357	$1,414	$751
Total	$51,341	$39,561	$29,103

From May (1994).
Amounts reported are the annual average in 1980 US$1000.

lost to agropastoral expansion represent an additional $29 to $51 million in annualized production foregone.

User Costs and Depreciation Values

The concept of sustainable income in the Hicksean framework is not directly consistent with that of sustainable production foregone. Rather, according to El Serafy's formulation (1988), it is necessary to determine the amount of rents that it would be necessary to retain in income-generating investments, so that, at the point at which the natural assets in question are exhausted, the accumulated asset base would be sufficiently large to generate an annual income, in perpetuity, equivalent to the sustainable production foregone. This amount may then be deducted from annual product as measured by the SNA, and the resulting net product may be considered as sustainable income.

TABLE 2.3 DEPLETION HORIZONS AND ANNUAL AVERAGE USER COSTS

Macroregion	*1971–1975*	*1976–1980*	*1981–1985*
		Depletion horizon (years)[a]	
North	888	363	451
Northeast	55	31	60
Center-West	84	55	195
Southeast	50	45	155
South	39	28	163
		User costs at 5% discount (1980 $1000)[b]	
North	$0	$2	$11
Northeast	$11,120	$46,239	$4,777
Center-West	$1,305	$8,695	$3
Southeast	$1,496	$4,809	$5
South	$6,614	$13,815	$3
Wood	$13,217	$45,827	$1,941
Nontimber	$8,435	$27,847	$2,854
Total, Brazil	$21,652	$73,674	$4,795

From May (1994).
[a]Timber resource depletion only.
[b]Includes user cost of nontimber forest products.

For this purpose, annual user cost estimates (U_{ij}) were derived for each Brazilian macroregion, applying the following equation:

$$U_{ij} = \frac{Y_{Ej}}{(1 + r)^n - 1}$$

where Y_{Ej} represents the sum of sustainable forest rents. Table 2.3 provides a summary of the average annual timber resource depletion horizons (n) and user cost estimates derived for census years in Brazilian macroregions for wood and nontimber products, applying a 5 percent discount rate for r.

The user cost concept is strongly related to two particularly sensitive and interrelated parameters: resource exhaustion horizon and the interest rate. For example, despite recent concerns, the large expanse of Amazon rainforest that remains even after accelerated clearing in the late 1970s effectively eliminates the value of this loss from the analysis, although sustainable production foregone due to Amazon deforestation was shown in table 2.2a to have been quite substantial,

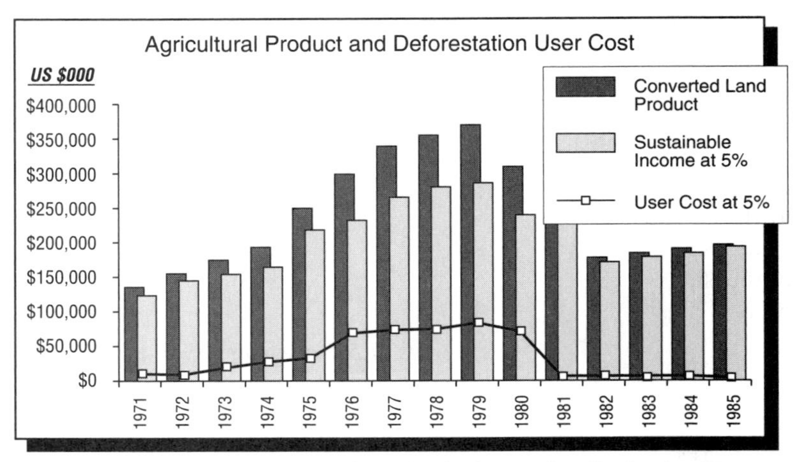

FIGURE 2.4 Sustainable Income from Agriculture, after Deducting Forest User Costs: 1970–85

rivaling only the rapidly growing frontier in the Center-West savanna for its forest losses. The loss is more significant still when we take into account the nontimber forest losses, which are in great measure confined to the Amazon.

Although in the period analyzed, the extent of deforestation of the Amazon forest had not yet reached alarming proportions, in other floristic zones the process of land conversion was already edging toward forest exhaustion by the late 1970s. Based on the period to resource exhaustion at the rate of depletion underway in the South and Northeast in 1975 to 1980, forest cover in these macroregions would be obliterated, respectively, in the years 2006 and 2009 (table 2.3). In the Southeast, a stabilization of depletion rates by the end of the 1970s pushed forest exhaustion out to the year 2023, but this conceals the fact that most reserves had already been removed by 1980.

To determine the sustainable income from agricultural activities that have supplanted forest land uses, the respective user cost values have been deducted from the estimated annual gross agricultural/livestock product (R_t) generated by farming the incremental agricultural lands (IBGE 1987). This procedure provides an estimate for sustainable income (gross agricultural product net of forest user costs, or $R_t - U_{ij}$). Annualized gross product, user cost, and sustainable income values are plotted in figure 2.4, whose source data are provided in table 2.4a.

TABLE 2.4a USER COST APPROACH

Year	*(A) User Cost (5% Discount)*	*(B) Converted Land Product*	*A/B*
1971	$11,904	$135,926	8.8%
1972	$8,394	$154,964	5.4%
1973	$18,821	$174,001	10.8%
1974	$28,352	$193,038	14.7%
1975	$32,371	$250,004	12.9%
1976	$68,274	$300,336	22.7%
1977	$74,187	$339,898	21.8%
1978	$73,030	$353,935	20.6%
1979	$82,154	$367,972	22.3%
1980	$70,723	$311,114	22.7%
1981	$5,293	$249,436	2.1%
1982	$4,754	$175,820	2.7%
1983	$4,869	$182,317	2.7%
1984	$5,376	$188,814	2.8%
1985	$3,737	$195,311	1.9%

From May (1994).
Amounts reported are in 1980 US$1000.
Sustainable agropastoral income = B − A.

In contrast, if one is to adopt the "net price approach" proposed by Repetto et al. (1989; Solárzano et al. 1992), net forest depreciation is equivalent to the entire net receipts that could be derived from marketing of forest products removed through deforestation and timber extraction. In this approach, the entire physical output of converted land is multiplied by the unit net receipt (rent) from wood harvest or loss. (There is no distinction in the net price calculation of the loss that would arise as a result of inability to sustainably harvest nontimber resources.) If such a procedure were adopted here, even assuming that only 25 percent of the value of all wood products available from converted forestland in Brazil could be treated as a net loss,[5] the results would be considerably different from those estimated by the user cost concept.

As can be seen in table 2.4b, a much larger value—in some years nearly equivalent to the entire agricultural product of Brazil—would then be considered as natural capital depreciation. In this light, the net price approach thus suggests that the nation would have been economically better off exploiting most of its forests for sustained

Table 2.4b Net Price Approach

Year	(C) Depreciation in Forest Stocks	(D) National Agricultural Product	C/D
1971	$6,074	$10,753	56%
1972	$6,582	$12,530	53%
1973	$8,220	$15,441	53%
1974	$13,145	$16,949	78%
1975	$16,334	$18,200	90%
1976	$15,985	$20,025	80%
1977	$16,819	$25,292	67%
1978	$14,411	$21,599	67%
1979	$13,766	$22,607	61%
1980	$11,296	$24,060	47%
1981	$9,111	$20,591	44%
1982	$6,558	$17,577	37%
1983	$6,336	$22,009	29%
1984	$6,081	$23,645	26%
1985	$5,161	$25,473	20%

From May (1994).
Amounts reported in 1980 US$million.

yield than in expanding its agricultural frontiers, whereas the user cost approach would identify those forests whose critical nearness to exhaustion indicates a higher cost of their substitution, but it would not oppose substitution in frontier areas.

CONCLUSIONS

This study has provided an empirical example comparing current approaches for adjusting national accounts to measure sustainable income. The net rent magnitudes measured according to the "net price" approach (a "strong sustainability" argument) tend to be significantly larger when compared with the user cost approach ("weak sustainability"). This is so because, instead of assuming prospects for reinvestment in other capital assets, the strong sustainability argument assumes that what is exploited today is gone for good. The net rent derived from resource removal is thus charged in its entirety against the sector responsible for the damage. As described, the net price valuation of merchantable timber lost as a result of deforestation in Brazil yields magnitudes approaching the entire gross national agricultural product. With the user cost approach, on the other hand, the values obtained for Brazil constituted less than half of 1 percent of the agricultural product.

To be sustainable, the income generated from wood and nontimber products must be sufficient to allow future generations to obtain the same welfare as the present generation. The user cost approach implicitly assumes that forests as assets can be replaced by other types of assets, i.e., material capital in other economic sectors with the same economic return. However, future generations are sure to derive many benefits from the retention of standing forests, that cannot be substituted by investments in other sectors equivalent only to the present value of sustainable on-site marketable forest products. Loss of off-site values of forests such as biodiversity and hydrologic and climate stabilization should also be considered as losses in future income from converted areas and, therefore, will increase the user cost and decrease the current value of agropastoral income from deforested areas.

To reflect the essential definition of limits to expansion of the scale of human activity, the restriction of areas conceptually "available" for agropastoral expansion must consider a range of factors, including land suitability, indigenous cultural values, and ecological resilience. That such recognition is possible was at least implied by the Brazilian government's "Green Package" of August 1996, when the proportion of agricultural establishments that could be cleared in the Amazon basin was reduced from 50 percent to 20 percent by a provisional measure.

The methods chosen for assessing movement toward sustainable development and appraising specific courses of action in this direction are far from neutral. The analytical framework colors the perception of what sustainability implies for the natural resource base. If indicators of sustainability help policy makers determine how much of a given ecosystem should remain intact to maintain vital physical survival functions and also economic well-being in the long term, they could be a tremendous boon to development decision making (King 1994). On the other hand, where agreement on the facts is impossible because of uncertainty, and where the relative desirability of substituting ecosystem functions for material well-being is controversial because of differences in peoples' underlying values, political negotiation and financial bargaining may be the only means to secure consensus.

In conclusion, the awareness that global common problems exist and are worsening is not enough to ensure a transition in the direction of sustainable development. Inequities in historical development processes make it difficult for Third World governments to allocate resources to conservation. This suggests the need to augment North-South transfers targeted toward stewardship of resources having transboundary significance (Martinez-Alier 1994). Such expression of "willingness to pay," to promote protection of resource uses that benefit the world at large and not only Brazil, could represent means to resolve issues of economic justice as well as of intertemporal equity.

APPENDIX

ANALYTICAL FRAMEWORK[6]

USER COST

This study follows the approach proposed by El Serafy (1988) to identify a "user cost" that represents the societal investment necessary in fixed capital that would ensure a sustainable income flow once forest resources are depleted. Such cost may then be considered to represent a loss that, in the national accounts, may be charged against receipts from those economic activities that have resulted in forestland conversion. The resulting net value represents the sustainable income from lands converted to agricultural use.[7]

To assess the evolution of renewable resource stocks in the presence of expanding agricultural frontiers, and their contribution to economic growth this study used the following analytical framework.

Let K_{j0} constitute the physical stock of a forest resource j at a particular point in time, and let g_j be the natural growth rate of that resource in terms of incremental volume or non–wood product output. Then K_T represents the stock of j in a second moment T, in the absence of extractive activity:

$$K_T = (1 + g_j)K_{j0}. \qquad (2.1)$$

Let Q_{jt} be the quantity of forest resource j that can be extracted in a period such that, at the end of the period, the forest stock is maintained at the same level as prior to extraction.

$$Q_{jt} = g_j \cdot K_{j0}. \qquad (2.2)$$

By definition, then, if p_{jt} is the net receipt from one unit of resource j, then it should be possible to ensure an infinite flow of income Y_{Ej} commensurate to a Hicksean "true income," provided extraction is maintained at a sustainable level.

$$Y_{Ej} = p_{jt} \cdot Q_{jt} \qquad (2.3)$$

However, activities underway in sector i result in the removal of V_{jt} ($\neq Q_{jt}$) of resource j in period t, obtaining R_{it} in net receipts. At this rate of forest conversion, the resource will be exhausted in n periods, calculated as $n = K_t / V_{jt}$.[8]

During the process of agricultural expansion, society bears an environmental loss as a result of forest conversion. To compensate for the effects of forest conversion to the national economy, a portion of the receipt R_i would have to be set aside as capital stock to ensure a sustainable income flow equivalent to Y_{Ej} once the forest resource stock is exhausted. This value, U_{ij}, known as "user cost," will now be derived for the case of forestland conversion for agriculture:

Let $F(U_{ij})$ be the future value of user cost accumulated during n periods, at a rate of interest equivalent to the opportunity cost of capital (r):

$$F(U_{ij}) = \sum_{t=0}^{n-1} (U_{ij})(1 + r)^t = (U_{ij}) \cdot \frac{(1 + r)^n - 1}{r}. \tag{2.4}$$

The present value $PV(U_{ij})$ of the accumulated capital reinvestment during this time horizon, using the social rate of time preference d as a discount rate, will be

$$PV(U_{ij})_t = F(U_{ij}) \cdot r \cdot \frac{1}{(1 + d)^t} = U_{ij} \cdot \frac{(1 + r)^n - 1}{(1 + d)^t}. \tag{2.5}$$

This last result may then be equated to the present value of the sustainable production Y_{Ej} that would be anticipated from the forest resource in the absence of land use conversion:

$$U_{ij} \cdot \frac{(1 + r)^n - 1}{(1 + d)^t} = \frac{Y_{Ej}}{(1 + d)^t}. \tag{2.6}$$

This expression simplifies finally to the following general statement of user cost:

$$U_{ij} = \frac{Y_{Ej}}{(1 + r)^n - 1}. \tag{2.7}$$

In summary, the values for forest resource loss estimation are based on the following variables:

U_{ij} = user cost of forest resource j attributed to activities in sector i
Y_{Ej} = value of wood and other forest products that could be removed in a sustainable fashion in a determined floristic zone

The remaining forest resource stock is estimated, sequentially, assuming uniformity in volumes and species composition for areas remaining in forest at the end of each period. The period n in years to resource exhaustion is then calculated, based on the forest removal rate registered in period t, as follows:

$$n_{jt} = K_{jt} / V_{jt},$$

where

K_{jt} = stock remaining of a given forest resource, and
V_{jt} = annual conversion of that resource for agricultural activities.

NET PRICE

The net price approach accounts for the physical variation of resource stock, multiplied by the market price of the product, net of production costs, and adjusted for price changes. This value is treated as equivalent to the rent accruing to the owners of the resource, representing the value of that portion of known stocks that decline as a result of extraction or conversion. This procedure deducts from total sectoral income the full estimated rent attributed to the resource depleted, as an indicator of resource depreciation, yielding a Net Product that reflects natural capital consumption.

The procedure takes account of additions to capital stock that result from discoveries, revisions, growth, or reproduction, as well as deductions caused by depletion, degradation, or deforestation. Price changes throughout the period in question are considered through revaluations of stocks and flows. The general formulation of this accounting identity is as follows:

$$X_{t+1}P_{t+1} = X_t p_t + (X_{t+1} - X_t)p^* + X_t(p_{t+1} - p_t) + (X_{t+1} - X_t)(p_{t+1} - p^*),$$

where

X_t = the opening stock of the resource in physical units,
X_t+1 = the closing stock of the resource in physical units,
p_t = the rent per physical unit at the opening of the period,
p_t+1 = the rent per physical unit at closing of the period, and
p^* = the average unit rent during the period.

This results in the following accounting relationships:

a. Net stock variation:

$$X_{t+1}P_{t+1} - Xtp = (X_{t+1} - X_t)p^* + X_t(p_{t+1} - p_t) + (X_{t+1} - X_t)(p_{t+1} - p^*)$$

b. Current net additions during the period:

$$(X_{t+1} - X_t)p^* = (Ad - Rd)p^*,$$

where

Ad = additions to stock (discoveries, net revisions, extensions, growth, and reproduction)
Rd = stock reductions (extraction, deforestation and degradation).

c. Revaluations:

$$Rv = X_t(p_{t+1} - p_t) + (X_{t+1} - X_t)(p_{t+1} - p^*),$$

where

$X_t\ (p_{t+1} - p_t)$ corresponds to the revaluation of opening stocks, and
$(X_{t+1} - X_t)\ (p_{t+1} - p^*)$ corresponds to the revaluation of transactions during the period.

NOTES

1. Turner (1994) discusses a gradation from "very weak" to "very strong" sustainability. The former was described by Daly and Cobb (1989) as referring to a "Disneyland effect" in which perfect substitution exists between natural and man-made capital, and society is forever equally happy as this substitution takes place. In the "very strong" version, Turner (1992) adds "cultural capital" to natural capital in defining limits to substitution. This would include the preservation of indigenous knowledge and cultural diversity.
2. The macroregions of Brazil are made up of the South-Central region, which includes the states of Rio de Janeiro, Espirito Santo, Minas Gerais and São Paulo; the South, which includes Paraná, Santa Catarina and Rio Grande do Sul; the Center-West, made up of Goiás, the Federal District, Mato Grosso and Mato Grosso do Sul; the Northeast, of Maranhão, Piauí, Ceará, Rio Grande do Norte, Paraíba, Pernambuco, Alagoas, Sergipe and Bahia; and the North or Amazon region, composed of Pará, Amazonas, Amapá, Roraima, Acre, Rondônia and Tocantins.
3. One additional species, the *piassava* palm, alone responsible for an additional 20 percent of nontimber product value, was not included in the exercise, because of the lack of information on resource productivity and sustainable extraction of fiber. However, this species occupies significant areas of the Atlantic Forest of Bahia, which has been subject to severe degradation during the period under study. The estimates for resource loss in nontimber resources are thus grossly underestimated.
4. An average wood price at point of export was obtained from CACEX data, dividing total revenues by volume of the full range of wood products shipped. The resulting price probably overstates average wood values, because a disproportionate amount of Brazilian wood exports consists of valuable hardwoods such as mahogany, of which 75 percent of national production is destined for export. Only a small share of total Brazilian wood production goes toward exports, however.
5. This proportion is similar to the effective relationship between potential timber harvesting and the actual value of wood obtained in the Amazon, where losses due to inefficient harvesting and milling techniques leave only about 25 percent of merchantable standing volume after extraction (see chapter 3).
6. This section draws heavily on Serôa da Motta and Young (1991:4–6).
7. This method has also been applied in studies of natural capital deprecia-

tion in the United States by Daly and Cobb (1990) and in environmental accounts for Mexico and Papua New Guinea prepared for the U. N. Statistical Office and the World Bank (van Tongeren et al. 1991; Bartelmus et al. 1992).

8. For the purposes of this study, we have assumed that forest growth rates are extremely small relative to the rate of forest land use conversion, characterizing deforestation as an analog to mining. This assumption is supported by the literature on primary and old growth secondary tropical forests whose increment is roughly balanced by mortality. This is not the case for young secondary forest and plantation stock, which have not been included in the present analysis.

3

Developing a Quantitative Framework for Sustainable Resource-Use Planning in the Brazilian Amazon

ORIANA TRINDADE DE ALMEIDA and CHRISTOPHER UHL

The Eastern Amazon has been aggressively settled since 1960 by loggers, ranchers, and farmers, among others. There have been dozens of books, symposia, and documentaries characterizing this settlement process (e.g., Moran 1981; Anderson 1990). Nevertheless, a consensus on how, or even whether, to occupy this vast region remains elusive. This is, in part, because there is a paucity of high-quality, objective analyses on the principal land-use activities leading to deforestation. Without a solid information base, conservationists, developers, and other interest groups pose arguments based on published statements that conform most closely to their own views. The net result is a polarized, slow-moving debate.

In an attempt to provide holistic and objective analyses of resource-use problems in the Eastern Amazon, researchers at IMAZON (*Instituto do Homem e Meio Ambiente da Amazonia,* or Amazon Institute of People and the Environment) have recently completed detailed analyses of logging, ranching, and farming in Paragominas Municipality, located along the Belém-Brasília Highway (respectively, Veríssimo et al. 1992; Mattos and Uhl 1994; Toniolo and Uhl 1995).

Results from these studies reveal that present-day land uses in the Eastern Amazon are frequently not sustainable from an ecological standpoint. Timber and soil nutrient resources are often mined with little thought for the future. This is a natural consequence of three factors: (1) land is abundant and cheap and, therefore, there is no incentive to be a good land steward; (2) capital is usually necessary to move from exploitative resource uses to sustainable uses, and settlers generally lack capital; and (3) sustainable resource uses require technical knowledge, and such expertise is rarely available in Amazonia.

Given these conditions, it is unlikely that sustainable approaches will be adopted voluntarily. Meanwhile, the human population is growing in the Eastern Amazon. In older settlement areas, land is no longer abundant. Hence, there is a growing need for the hand of government to coax and guide resource users toward more responsible behaviors.

In Brazil, as elsewhere, government is charged with looking after the well-being of the environment and defending the rights of all citizens both born and unborn. In 1988, Brazil adopted a new constitution that accords significant power to municipal governments. These local governments are now well positioned to promote the wise and sustainable use of Amazonian resources.

Our objective in this chapter is to synthesize results from our previous economic and ecological studies in Paragominas and show how they can be used to promote sustainable regional planning there and elsewhere in the Eastern Amazon. The chapter is divided into three sections. First, we compare land-use activities (logging, ranching, and farming) in their extensive and intensive forms using economic criteria (gross return, profit, tax generation), social considerations (employment generation), and an ecological parameter (carbon liberation). Second, we scale up to the municipal level to provide estimates of the gross and net returns from ranching, farming, and wood industries for Paragominas. In this section, we also forecast declines in returns in the near future that would result from the nonsustainable nature of present land-use practices, and we use our data to consider alternative, sustainable, resource-use models for this municipality. Finally, we discuss the economic, political, and legislative tools available for promoting sustainable resource uses at the municipal level in the Brazilian Amazon. The common thread running through all these analyses is that high-quality information has an important, yet hitherto largely unexplored, role to play in guiding natural resource use in Amazonia and elsewhere in the developing world.

GENERAL CHARACTERISTICS OF LAND USE IN PARAGOMINAS

Paragominas Municipality, founded in 1965, contains 2.5 million hectares (ha) and a population of 92,355 people (FIBGE 1991).[1] This municipality presents a microcosm of Eastern Amazonia, containing within its boundaries significant areas devoted to extensive-style ranching, logging, and slash-and-burn agriculture. By 1988, 34 percent of the municipality had been deforested or logged. At that time, there were 1.6 million ha of mature forest, 242,000 ha of logged forest, 263,000 ha of young second-growth forest, and 352,000 ha dedicated to ranching and agriculture (Watrin and Rocha 1992; figure 3.1).

The three groups—farmers, ranchers, and loggers—arrived in Paragominas at different times, and each group has had distinct economic and ecological impacts on the region. Shifting cultivation (figure 3.2a) was the first activity to be practiced by modern settlers. Farmers first arrived in the 1930s, and gradually small farm colonies formed. Today, there are twenty farming colonies in Paragominas, each occupying about 2,500 ha. Although most colonists still plant rice, cassava, and corn using extensive, shifting-cultivation methods, a few use more intensive approaches, cultivating perennials, such as black pepper, oranges, and passion fruit (figure 3.2b).

Ranchers arrived in Paragominas in the 1960s with the opening of the Belém-Brasília Highway. They made large pasture clearings to lay claim to the land. Most clearings were poorly managed and poorly stocked (i.e., <0.5 animal/ha). Weeds invaded many pastures, forcing abandonment (Serrão and Falesi 1977). Other areas are still being grazed today following this extensive model (figure 3.2c).

Starting in the 1980s, a few ranchers began to rejuvenate their abandoned pasture lands by bulldozing away brush and debris, tilling the soil, fertilizing, and sowing better-adapted forages (e.g., *Brachiaria brizantha*). This more intensive approach (figure 3.2d) allows for a higher stocking density (1 animal/ha) and greater productivity. With periodic fertilization (every five years), this system appears to offer promise (Mattos and Uhl 1994).

Logging is the third important land-use activity in Paragominas. In 1970, when ranching was the dominant economic activity, only one sawmill was present in Paragominas. However, by the advent of the 1990s, there were 137 mills operating there (Veríssimo et al. 1992). Most mills consist of a single band saw and produce standard-dimension boards.

Mill owners frequently obtain timber by purchasing the rights to log forest tracts owned by ranchers. Timber is extracted in a chaotic fashion using heavy machinery; four to eight trees are removed per hectare (equivalent to 25–50 m^3/ha; figure 3.2e). Approximately 4 m^3 of wood are unnecessarily destroyed for each m^3 of sawn wood that is produced. The cutting cycle required to sustain this practice of extensive-style logging is ninety years (Barreto et al. 1993).

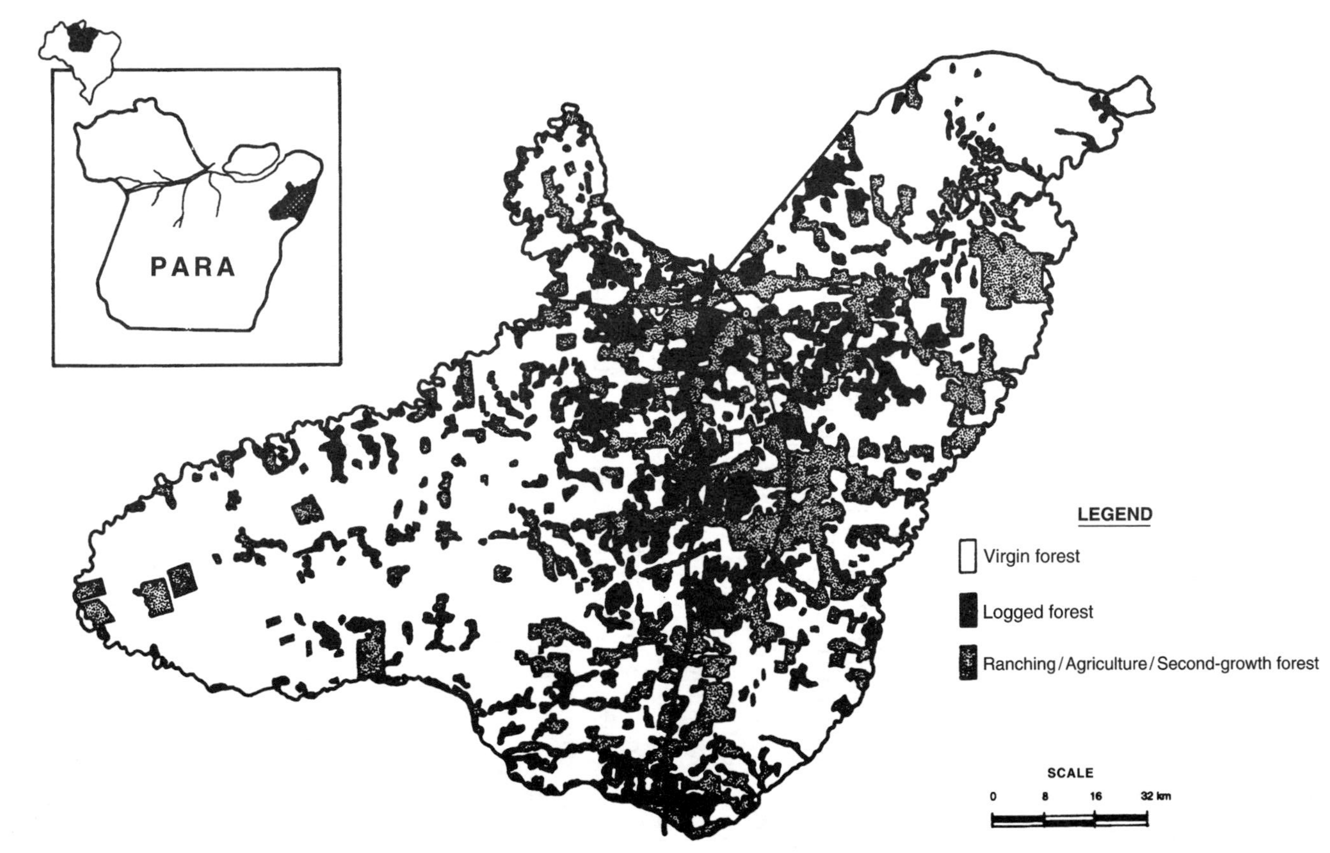

FIGURE 3.1 Paragominas Municipality, Showing Virgin Forest Areas, Areas Subjected to Extensive-style Logging, and Areas Cleared for Ranching and Agriculture *Source:* Watrin and Rocha (1992).

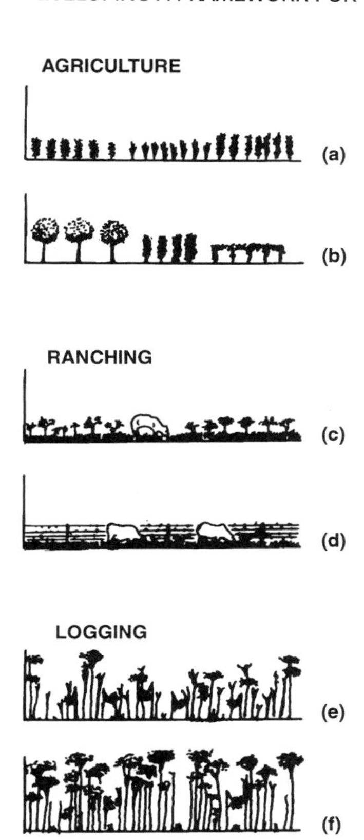

FIGURE 3.2 Schematic Representations of the Extensive and Intensive Approaches to Farming, Ranching, and Logging in the Eastern Amazon ***Note:*** Most farming is done using the extensive-style slash-and-burn technique (a), but permanent cropping with perennials (b) is also possible. Ranches typically involve large clearings and low herd densities (c), but beef-cattle ranching on improved pastures (d) is possible. Typical logging operations entail the removal of 4 to 8 trees/ha, but much damage is incurred in these unplanned extraction episodes (e); simple forest management measures (f), such as preharvest vine cutting, directional felling, skid trail planning, and postharvest ringing of defective trees, would substantially increase productivity and reduce cutting cycle length.

It would be possible to both plan timber extraction to reduce logging damage and apply postharvest silvicultural treatments to increase tree growth. According to analyses by Barreto et al. (1993), this more intensive, "management" approach to forestry (figure 3.2f) would reduce the cutting cycle (the interval between harvests) from ninety to approximately thirty years.

At present, approximately 80 percent of the land surface of Paragominas is controlled by ranchers; loggers own some 16 percent of the municipality (M. Uzeda and C. Uhl, IMAZON, internal doc.), and 3 to 4 percent is in the hands of small farmers. Ranchers are increasingly involved in logging as they sell logging rights to their forest tracts or conduct logging themselves. At the same time, sawmill owners are increasingly inclined to invest surplus capital in forest land, ensuring themselves a stock of timber in the future.

It should be borne in mind that settlement activities started earlier in Paragominas than in most other areas of the Eastern Amazon. Because of its more mature status, Paragominas serves as a bellwether municipality: development trends, innovations, and failures frequently appear in Paragominas before they are seen in other, younger development zones. It follows that lessons learned in Paragominas could have broad import in influencing regional development.

METHODOLOGY

In the early 1990s, researchers at IMAZON engaged in detailed studies to evaluate the economic, ecological, and social characteristics of logging, ranching, and farming in Paragominas. We use the data from these and other studies to compare both extensive and intensive approaches to these land uses in terms of their gross returns, their profits, the taxes and employment they generate, their capital requirements, and their associated environmental implications.

A study by Veríssimo et al. (1992) provides information on logging in Paragominas, including data on profits and costs of extraction, transport, and wood processing. We use this study to characterize extensive-style logging and wood-processing activities. Data on costs and returns from intensive, management-based timber operations come from a recent IMAZON study by Barreto et al. (1993). Data on gross returns and profits for both extensive-style ranching and more intensive "improved" ranching are from an IMAZON study by Mattos and Uhl (1994). Finally, data on net returns and profits for extensive-style shifting cultivation and intensive permanent cropping are from a recent IMAZON study by Toniolo and Uhl (1995) and from studies by Ribeiro (1989), Albuquerque et al. (1989), and records from SAGRI (Pará's Secretary of Agriculture).

We estimated the capital required to initiate each land-use activity. For wood companies, start-up costs included the cost of machinery, equipment, vehicles, and buildings. For ranching, we considered the cost of fences, corrals, salt licks, roads, houses, vehicles, and animals. Our estimate of the capital necessary to undertake slash-and-burn agriculture included labor to clear the forest and plant crops, and the cost of seed. For intensive agriculture, which usually involves the growing of perennial crops, start-up costs included the price of labor to clear and plant the land and capital for the purchase of seedlings, fertilizers, and other inputs.

Taxes were assessed based on the gross revenues from each activity. The ICMS (Interstate and Intermunicipal Product Transport Tax) is the principal tax collected on wood and agricultural products. The ICMS differs depending on the product and its destination and is assessed as follows: 17 percent of gross returns for products sold within the state of Pará (e.g., beef, oranges, passion fruit, cassava flour, and corn), 12 percent for products sold in other Brazilian states (wood products), and 13 percent for products that are exported (e.g., black pepper).

Using these data, we calculated the costs and revenues associated with extensive and intensive forms of logging, ranching, and agriculture. To determine the net present value and internal rate of return for each activity, we first used annual discount rates of 6 percent. A discount rate of 6 percent is considered the minimum rate acceptable to investors; this is the rate provided by savings banks in Brazil. Sophisticated investors can achieve returns higher than 6 percent because of capital scarcity and speculation. Hence, we also conducted the analysis using a 12 percent rate; this rate reflects what an investor with access to market information might receive. In recent years returns on market investments have sometimes been less than 12 percent (e.g., 1991) and have sometimes exceeded this value (e.g., 1992 and 1993). We used a time frame of ninety years for our analyses because this is the estimated interval between timber harvests for natural forest in the absence of management.

We estimated the loss of carbon that results from the conversion of forest for each land use and treated the result as one indicator of the ecological cost of each activity. Carbon loss is considered to be proportional to the decline in biomass that results from forest conversion. Our estimates of biomass loss are based on data found in Uhl et al. (1988), Veríssimo et al. (1992), DeGraaf (1986), Nepstad (1989), and Barreto et al. (1993).

RESULTS AND DISCUSSION

Annual Operating Costs and Returns for Land-Use Activities

In the analysis that follows, we compare extensive and intensive approaches to logging, ranching, and agriculture in the Eastern Amazon. In all cases, we conduct the analysis stipulating that the activities be practiced in an environmentally sustainable fashion. By this we mean that the structure, nutrient content, and nutrient and water holding capacity of the soil not deteriorate over time. All results ae expressed on an annual basis considering one hectare and therefore show what a practioner might actually spend and receive on an annual basis.

Timber Extraction and Processing

The extensive-style logging operations characteristic of the Paragominas region removed, on average, 38 m^3/ha of roundwood (Veríssimo et al. 1992). A typical mill required, on average, 9,200 m^3 of roundwood/year. Hence, each mill harvested timber from approximately 242 ha/yr (9,200 m^3/38 m^3). Given a cutting cycle of ninety years (required for sustainable practice), a typical mill would need about 21,780 ha (242 ha/yr $\times$ 90 yr).

In this system, the total value of the wood harvested from 1 ha (38 m^3) after sawing was \$2,772 (table 3.1, note b). Hence, the gross return per hectare over the ninety-year cycle would be \$31/yr (\$2,772/ 90 yr). Annual tax generation, therefore, was estimated at \$4/ha (0.12 $\times$ \$31) (table 3.1).

The profit, considering the entire 21,780 ha area required for a sustainable operation, would be \$11/ha/yr. The investment to establish a logging and milling operation was approximately \$2,391/ha: 36 percent of this investment was for extraction equipment, 33 percent was for log transport trucks, and 31 percent was for construction of the sawmill (Veríssimo et al. 1992). These investment costs could be paid off in about 2.5 years (\$2,391/\$961; table 3.1, note c).

An alternative to this unplanned approach to timber resource use is forest management (column 2, table 3.1). With management, the extraction of wood is carefully planned and silvicultural treatments are applied following timber extraction to accelerate the growth of the commercial trees remaining in the stand. These measures are projected to reduce the cutting cycle from ninety to approximately thirty years while ensuring the accumulation of at least 1.2 m^3/ha/yr of com-

TABLE 3.1 FINANCIAL VIABILITY OF EXTENSIVE AND MANAGED WOOD INDUSTRY IN PARAGOMINAS

	Prevailing Extensive Model	*Alternative Intensive Model*
	(Typical Company; 90-Year Cutting Cycle)[a]	*(Management-Based Company; 30-Year Cutting Cycle)*[a]
Gross return ($/ha/yr)[b]	31	92
Profit ($/ha/yr)	11[c]	28[d]
Start-up investment ($/ha)	2,391[e]	2,504[f]
Tax generation ($/ha/yr)[g]	4	11

[a]These data are summarized from Veríssimo et al. (1992) and Barreto et al. (1993). A 90-year cutting cycle is necessary for sustainable forestry in the extensive model; a 30-year cycle is presumed for the management-based approach. We just consider annual costs in this analysis; the discount rate is 0%; the cost of capital is included in the long-term analysis (see table 3.4).
[b]The gross return for a typical wood company was $670,800/yr (Veríssimo et al. 1992). The average area harvested per mill per year was 242 ha. Therefore, the average gross return per hectare was $2,772 ($670,800/242 ha) and the return/ha/yr was $31 ($2,772/90 years).
[c]The total cost of extensive-style extraction and processing, including depreciation and taxes, was $438,239 for a typical wood company, or $1,811/ha ($438,239/242 ha). Therefore, the profit for a typical wood company was $961/ha ($2,772 − $1,811), or $11/ha/yr ($961/90 yr). Note: We used a somewhat higher "maintenance cost" than Veríssimo et al. (1992), based on additional information from J. Zweede.
[d]The costs for management-based logging were the same as for extensive-style logging, with the addition of a management cost of $113/ha. In this case, the profit would be $848/ha/30 yr ($961 − $113), or $28/ha/yr.
[e]The start-up investment to establish a complete logging and milling operation that processes 9,200 m^3 of roundwood per year (typical capacity of Paragominas mill with 1 bandsaw) would include the purchase of 2 chain saws ($700 each), 1 bulldozer ($120,000), 2 log loaders ($90,000 each), 3 log-transport trucks ($65,000 each), 1 bandsaw ($42,000), 1 sawblade-sharpening machine ($9,900), 1 table saw ($2,300), 1 log turner ($3,000), and 1 mill shed with office space ($25,000), for a total investment of $578,600 (Veríssimo et al. 1992). Considering a 90-year cutting cycle, this would be equivalent to a one-time investment of $2,391/ha ($578,600/242 ha.
[f]The initial investment for a wood industry with management was the same as for one without management ($2,391/ha; see previous footnote), with the addition of $113/ha for forest management costs. The total investment was $2,504/ha ($2,391 + $113).
[g]Products sold outside of Pará were charged a tax (ICMS) equivalent to 12 percent of gross returns.

mercial wood (Barreto et al. 1993; DeGraaf 1986). Gross returns and profits under forest management are estimated at $92/ha/yr and $28/ha/yr, respectively (table 3.1). Tax generation would be $11/ha/yr.

The money needed to implement forest management operations would be $2,504/ha. Although this forest management approach would involve higher costs and a reduction in profit margins, the cut-

TABLE 3.2 FINANCIAL VIABILITY OF EXTENSIVE AND IMPROVED BEEF-CATTLE RANCHING IN PARAGOMINAS

	Prevailing Extensive Model	*Alternative Intensive Model*
	(Unimproved Pastures[a])	*(Improved Pastures[a])*
Gross return ($/ha/yr)	31	104
Profit ($/ha/yr)	2–6[b]	55
Start-up investment ($/ha)	307[c]	539[d]
Tax generation ($/ha/yr)[e]	5	18

[a]Estimates are based on data from Mattos and Uhl (1994). Discount rate is 0 percent.
[b]Fertilization might be necessary to maintain productivity on extensive-style pastures. In this case, returns are estimated at $2/ha/yr. Should fertilization not be required, returns would be $6/ha/yr.
[c]Start-up investments per hectare for a typical extensive-style, beef-cattle ranch at Paragominas included infrastructure ($63), herd purchase ($134), pasture formation ($81), and vehicles and work animals purchase ($28). The total investment was $307/ha.
[d]Capital investments to restore degraded pastures were $260/ha. Beyond this, investments in livestock ($252/ha) and infrastructure, vehicles, and work animals ($27/ha) were necessary. The total capital investment to establish an improved pasture was $539/ha ($260 + $252 + $27).
[e]Products consumed within state boundaries are charged a tax (ICMS) equivalent to 17 percent of gross returns.

ting cycle would be reduced by two-thirds. Consequently, we estimate that revenues, profits, taxes, and jobs would all increase (table 3.1).

Ranching

The prevalent, extensive-style approach to ranching in Paragominas had a gross return of $31/ha/yr. Profits, excluding fertilizer as a cost, were $6/ha/yr (Mattos and Uhl 1994), but to place extensive-style ranching on sustainable footing, we consider that fields must be fertilized every fifth year at a cost of $20/ha (equivalent to an average outlay of $4/ha/yr). Including these costs, profits from ranching drop to approximately $2/ha/yr.

The capital necessary to establish an extensive-style ranch was estimated at $307/ha, divided among costs to purchase the herd (44 percent), establish pastures (26 percent), build infrastructure (fencing, corrals, salt trays, roads, houses), and purchase work animals (Mattos and Uhl 1994; table 3.2).

The rejuvenation of pasture in weed-infested abandoned paddocks represents a move toward intensification. Reforming a pasture consists of tilling, fertilizing, and planting the hardy forage *Brachiaria brizantha,* followed by periodic applications of fertilizers. A rejuvenated pasture supports more animals (1 head/ha instead of 0.5 head/ha) and the weight gain of the animals is higher (Mattos and Uhl 1994). The cost to reform and stock one hectare of pasture is $539/ha: 48 percent of this cost is for the reform itself, 47 percent is to stock the pasture at 1 head/ha, and the remaining 5 percent goes toward improving infrastructure and labor costs. The gross return per hectare on these "improved" pastures averages $104/ha/yr, profit is $55/ha/yr, and tax generation is $18/ha/yr (table 3.2).

Agriculture

Traditional, extensive-style agriculture involves the simultaneous cultivation of cassava, rice, and corn using slash-and-burn techniques. Fields are usually farmed for one to two years. To ensure environmental sustainability, the recommended fallow period between plantings is ten years. The gross return per planting cycle (i.e., eleven to twelve years) was $90/ha, when expressed on an annual basis ($1,079/12 yr), and the profit was $33/ha/yr (table 3.3). Tax generation was $15/ha/yr. Approximately $290/ha of capital or its labor equivalent was required to establish a slash-and-burn farm (table 3.3).

In the more intensive model of agriculture, involving the planting of perennials, inputs of fertilizers and pesticides are required. The gross returns from one hectare of land that was planted in equal proportions with oranges trees, black pepper vines, and passion fruit vines were estimated at $2,366/yr. Profits were estimated at $802/ha/yr and tax generation was $367/ha/yr. However, considerable capital was required to establish these systems: Start-up costs were $2,695/ha, 76 percent for materials and the rest for labor.

Summary

The more intensive land-use approaches in Paragominas consistently resulted in greater gross returns, taxes, and profits per hectare, based on this analysis (i.e., using 0 percent discount rate and just considering annual operating costs and revenues; tables 3.1, 3.2, and 3.3). From the point of view of local governments, these intensive approaches are desirable. Management-based wood industries and improved ranching practices resulted in a threefold or greater increase in gross returns and taxes over the extensive forms of these same ac-

TABLE 3.3 FINANCIAL VIABILITY OF EXTENSIVE SLASH-AND-BURN AGRICULTURE AND INTENSIVE PERENNIAL CROPPING IN PARAGOMINAS

	Prevailing Extensive Model	*Alternative Intensive Model*
	(Slash-and-Burn: Annual Crops[a])	*(Intensive Agriculture: Perennial Crops[a])*
Gross return ($/ha/yr)	90[b]	2,366[c]
Profit ($/ha/yr)	33[b]	802[d]
Start-up investment ($/ha)	291[e]	2,695[f]
Tax generation ($/ha/yr)	15[g]	367[h]

[a]The products produced in slash-and-burn agriculture are rice, corn, and cassava. Data are from Toniolo and Uhl (1995), Albuquerque et al. (1989), Ribeiro (1989), and SAGRI (unpubl.). Figures reflect projected outcomes when the activities are practiced in an environmentally sustainable fashion. Discount rate is 0 percent. All values were corrected for inflation and expressed in 1992 dollars.

[b]Toniolo and Uhl (1995) report that gross returns were $1,079 during one cropping cycle. To practice shifting cultivation sustainably, we consider that a 10-year fallow must occur between the cropping episodes that last 1 to 2 years. Hence we divide gross and net returns by 12 to express results on an annual basis.

[c]To calculate the gross returns from perennial-based agriculture, we considered that 1 ha is planted in equal proportions of orange trees, black pepper vines, and passion fruit vines (the most frequently planted perennials at Paragominas). The average annual returns of 1 ha planted to orange trees was $2,700 (Ribeiro 1989), 1 ha planted to black pepper vines, $2,665 (Albuquerque et al. 1989), and 1 ha planted to passion fruit vines, $1,733 (SAGRI, internal document), for an average of $2,366/ha/yr.

[d]The annual profits from 1 ha planted with orange trees was $1,627, 1 ha planted with black pepper vines, $409, and 1 ha planted to passion fruit vines, $369. Considering equal parts of 1 ha planted to each of these cultivars, the average annual profit is estimated at $802/ha.

[e]The start-up investment to establish a one-hectare slash-and-burn plot is estimated at $291, and includes the cost of rice and corn seeds ($11) and the labor to cut, burn, plant, weed, and harvest the land ($280).

[f]The start-up investment to establish a one-hectare plantation of orange trees was estimated at $1,897 (Ribeiro 1989), 1 ha of black pepper vines, $4,269 (Albuquerque et al. 1989), and 1 ha of passion fruit vines, $1,920 (SAGRI, internal document). Considering equal portions of a hectare planted to each of these cultivars, the average capital investment was $2,695/ha.

[g]Products consumed in Pará are charged a tax (ICMS) equivalent to 17 percent of gross returns.

[h]Black pepper is an export crop and is taxed at 13 percent; passion fruits and oranges are consumed in Pará and, therefore, taxed at 17 percent. Taxing gross returns of each of these crops (see footnote b) at these rates, generates an average of $367/ha/yr in ICMS revenues.

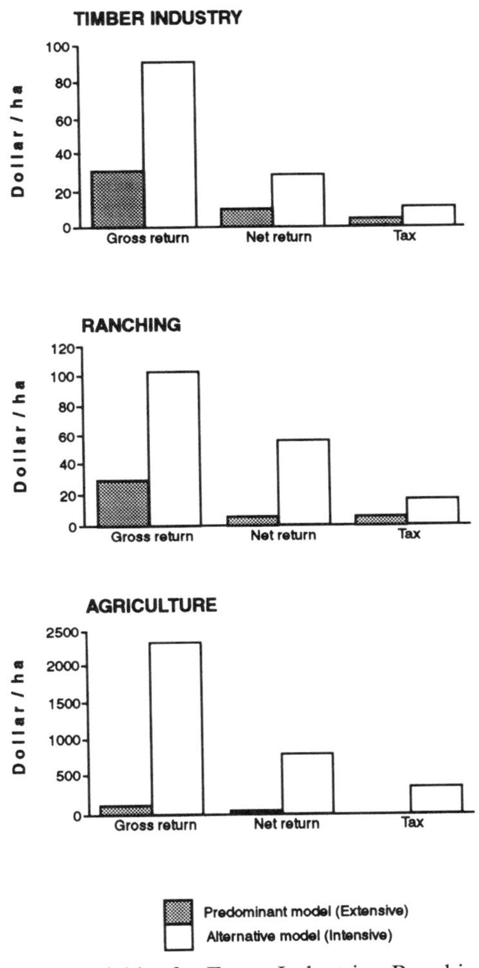

FIGURE 3.3 Financial Variables for Forest Industries, Ranching, and Agriculture Following Extensive and Intensive Practices, Paragominas

tivities (figure 3.3). Revenues from perennial agriculture were 26 times higher than from slash-and-burn agriculture, and taxes were projected to be 24 times higher.

From the point of view of the individual land holder, we observed that profits also increased when the practice of these three activities was intensified: profits from the wood industries increased from $11 to $28/ha/yr; ranching, $26 to $55/ha/yr; and agriculture, $33 to $802/

ha/yr (figure 3.3). However, intensive land-uses also have higher capital requirements. For example, ranching and agriculture require 2 to 10 times more start-up capital than their extensive-style counterparts.

Long-term Economic Performance of Land-Use Activities

To compare these land-use approaches, it is important to consider the returns per unit land area in a given year, as we have done (tables 3.1, 3.2, and 3.3), but it is also useful to include the start-up costs of the activity and the costs of having capital tied up over long time periods. In table 3.4, we compare the long-term performance of logging, ranching, and agriculture in both their intensive and extensive forms. Specifically, we calculated the net present value (NPV) at 6 percent and 12 percent discount rates, and the internal rate of return (IRR) over a ninety-year period.

The NPV of timbering operations when logging rights are purchased is $164/ha (6 percent discount rate), and the IRR is 108 percent. If the company were to purchase its own land for logging, the NPV would fall to $36/ha, with an IRR of 8 percent (table 3.4). When forest management is implemented, the volume of timber removed triples, since timber is harvested every thirty years instead of every ninety years. The NPV for this scenario shoots up to $459/ha when extraction rights are purchased (compared to $164/ha without management). However, management is more likely to occur on the logging firm's own land. If the company purchased its own forest land, the NPV of timber harvesting and processing would be $318/ha, a value still considerably greater than the nonmanagement case.

Profits are lower for company owners who only extract, and do not process, timber. In this case, the NPV of harvesting timber on somebody else's land, without practicing forest management, would be $35/ha, and the IRR, 29 percent. This approach is not viable for a firm that harvests timber from its own land (NPV = *negative* $93/ha; IRR = 2 percent).

Overall, this analysis of timber industries demonstrates that the returns on forest management investments are small when the company owns its own forest land (5 to 16 percent; table 3.4), in large part because there is a thirty-year lag between the investment in forest land and the return from this investment. Although the situation improves when management is done without purchasing the land, this is an unlikely scenario. Given that forest management for timber production on one's own land has a relatively high cost compared to other industry options, and recognizing that forest management techniques are poorly understood by most landholders, there is little inclination to invest in forest management.

To compare ranching with wood industries, we considered the initial investments and ongoing capital requirements of ranching over a ninety-year cycle. Since pastures in extensive ranching systems are usually degraded within the first ten years of use, we included the additional cost of applying fertilizer every fifth year to keep these pastures productive for the full ninety-year cycle (see Methods). In this extensive ranching model, the NPV over ninety years is *negative* $285/ha (6 percent discount rate), which shows that extensive-style cattle ranching is probably not viable over the long term (see also Mattos and Uhl 1994). Reforming pastures improves this scenario significantly by increasing the NPV to $516/ha with a 13 percent IRR.

Agriculture generates by far the highest NPV per hectare over a ninety-year period (table 3.4). Extensive-style, slash-and-burn agriculture has an NPV of $648/ha (6 percent discount rate), and perennial agriculture has an NPV of $13,502/ha, with a 70 percent IRR. However, certain factors may prevent farmers from realizing such high returns. For example, the market for agricultural products is subject to sudden fluctuations that can result in significant losses for farmers. This occurred in 1992, when the price of passion fruit fell 50 percent, from $0.22/kg (price in 1991) to $0.11/kg.

Considering a 12 percent discount rate, intensification becomes decidedly unattractive for many activities. Both extensive- and intensive-style ranching have negative NPVs (table 3.4). And in the case of wood industries that buy their land, three of four scenarios have negative NPVs. In the case of agriculture, at a 12 percent discount rate, NPV falls to $384/ha for extensive practices and $6,049/ha for intensive practices.

Social Considerations: Employment Generation

The principal land-use activities in Paragominas differ in their capacity to directly generate employment. Typical, extensive-style wood industries have a low labor requirement when expressed on an areal basis. Considering a ninety-year harvest cycle (sustainable extensive model), 531 ha of forest are required to provide employment for one person (table 3.5). In the case of a wood industry that practices forest management (harvest cycle reduced to thirty years), employment generation would more than triple: Each 154 ha of managed forest would provide one job (table 3.5).

TABLE 3.4 FINANCIAL INDICATORS FOR WOOD INDUSTRIES, RANCHERS, AND FARMERS UNDER DIFFERENT LAND-USE MODELS IN PARAGOMINAS[a]

Land-Use Models	*Net Present Value[b] ($/ha)*		*Internal Rate of Return[b]*
	Discount Rate		
	6%	*12%*	
Wood industries with no investment in land[c]			
Extraction and processing (w/o management)	$164	$75	108%
Extaction and processing (with management)	$459	$189	103%
Extraction only (w/o management)	$35	$12	29%
Extraction only (with management)	$145	$49	33%
Wood Industries that Own the Land[d]			
Extraction and processing (w/o management)	$36	−$52	8%
Extraction and processing (with management)	$318	$55	16%

Extraction only (w/o management)	−$93	−$115	2%
Extraction only (with management)	−$36	−$123	5%
Ranching			
Extensive	−$285	−$279	−0.1%
Intensive (improved pastures)	$516	−$8	13%
Agriculture			
Extensive (slash and burn farming)	$648	$384	236%
Intensive (perennial cropping)[e]	$13,502	$6,049	70%

[a]This analysis is performed considering a use-period of 90 years (the presumed forest cutting cycle in the absence of forest management) for all activities. Start-up costs are summarized in tables 3.1, 3.2, and 3.3 and are based on Veríssimo et al. (1992), Mattos and Uhl (1994), Toniolo and Uhl (1995), and Barreto et al. (1993).

[b]Definitions: Net present value = $\sum_{t=0}^{n} [(B_t - C_t)/(1 + i)^t]$; internal rate of return finds i, such that $\sum_{t=0}^{n} [(B_t - C_t)/(1 + i)^t] = 0$, where B_t is benefits in each year t; C_t is costs in each year t, n is number of years to the end of the project, and i is the discount rate.

[c]Logging rights are purchased for $70/ha.

[d]We consider that the price of forest land is $150/ha (typical price in 1993). When logging is conducted without management, we consider that 21,780 ha must be purchased to sustain a typical mill (242 ha/yr × 90 yr). With management, the mill would need one-third of this area (242 ha × 30 yr = 7,260 ha).

[e]We consider the lifetime of a passion fruit plantation to be 3 years, a black pepper plantation, 12 years, and an orange orchard, 10 years. Hence, in our analysis, the investments to establish the passion fruit plantation are repeated every 3 years and to establish pepper vine and orange tree plantations, every 12 and 10 years, respectively.

TABLE 3.5 EMPLOYMENT GENERATION POTENTIAL OF WOOD INDUSTRIES, RANCHING, AND AGRICULTURE OPTIONS IN PARAGOMINAS

Land-Use Activities	*Hectares to Employ One Person*
Timber Harvest and Processing	
Prevailing model: extensive/traditional wood industry	531[a]
Alternative: intensive/management-based industry	154[b]
Ranching	
Prevailing model: extensive/unimproved pastures	95[c]
Alternative: intensive/improved pastures	45[c]
Agriculture	
Prevailing model: extensive/slash-and-burn agriculture	16[d]
Alternative: intensive/perennial cropping	1.4[e]

[a]The prevalent extensive-style wood industries have an extraction crew of 10, a log transport crew of 3, and a mill crew of 28. This group of workers harvests and processes wood from 242 hectares/year (Veríssimo et al. 1992). Hence, 5.9 ha (242 ha/41 workers) is required to employ each person in a given year. Considering a 90-year harvest cycle, we calculate that 531 ha provide employment for one person in such extensive-style wood industries.

[b]In management-based wood industries, the harvest cycle is one-third that for extensive-style industries. This alone reduces the area per worker from 531 ha to 177 ha (531/3). Furthermore, in management-based forestry, people are employed to plan the logging and to conduct silvicultural treatments. A total of 39.5 hours of work are required to manage 1 ha of forest for timber production or 1.32 hr/ha/yr (39.5 hr/30 yr). A management-based timber company harvests 242 ha/yr and requires a total holding of 7,260 ha (242 ha × 30 yr) for a sustainable operation. If 1.32 hr are required per hectare per year for management, then a total of 9,583 hours are required for the entire area each year (1.32 hr × 7,260 ha). Considering a 6-hr effective field work day, and 262 work days/yr, 6.1 people would be required for management [9,583 hours/(262 days × 6 hours)]. The total staff requirement to manage and process the wood production from the 7,260-ha area would be 10 harvesters + 3 log transporters + 28 mill workers + 6 forest managers, or 47 people. Hence, each 154 ha would provide employment for one person (7,260 ha/47 people).

[c]Extensive-style ranching requires approximately 95 ha to generate each job (Mattos and Uhl 1994). Information on labor practices for improved pastures is limited, but labor needs are greater because (1) greater attention is given to animal health, (2) paddocks are smaller and herd rotation is more frequent, and (3) more infrastructure (fences, corrals) are present and more effort is given to the maintenance of this infrastructure. We assume here that labor requirements per unit area double in intensive-style ranches.

[d]Extensive-style slash-and-burn agriculture requires 189 days of labor per hectare for all activities. Given a work year of 262 days, 1 person can farm 1.33 ha (262 days/189 days).

Considering a 12-year rotation cycle, 16 ha are required to employ one person (1.33 ha/yr × 12 yr).
[e]The average labor, including time required to establish, maintain, and harvest crops is 105 days/ha/yr for orange cultivation, 342 days/ha/yr for black pepper, and 103 days/ha/yr for passion fruit (Ribeiro 1989, Albuquerque et al. 1989, SAGRI, internal document). If 1 ha is divided into equal plantings of each, the average labor is 183 days/ha/yr. Given 262 work days/yr, 1.4 ha are required to employ one person in this perennial cropping model (262 days/183 days).

Extensive-style ranching employs one person for each 95 ha of pasture. Most of this work force consists of temporary, day laborers. Production per hectare goes up when abandoned pastures are rejuvenated and so do labor requirements: We suppose that one job is created for every 45 ha in the case of ranching using improved pastures and herd management practices.

In traditional, slash-and-burn agriculture, one person is able to perform all farm tasks in a 1.33 ha area. Hence, 16 ha are needed to support one person, given the twelve-year cycle of planting and fallows (1.33 ha × 12 yr). Permanent agriculture with perennial crops is very labor intensive, requiring one person for each 1.4 ha in use (table 3.5).

Environmental Considerations: Carbon Emissions

The liberation of carbon dioxide through deforestation is contributing to changes in global climate. Reis and Margulis (1991) estimated that 15 to 30 percent of global CO_2 liberation resulted from deforestation (1–2 gigatons/yr), and that deforestation in Amazonia alone might be responsible for 4 to 6 percent of the CO_2 released annually to the atmosphere. One way to evaluate the amount of carbon released from deforestation is to estimate the difference between the amount of carbon stored in a mature forest and the amount of this biomass that is lost as a result of a given land-use. We did this for logging, ranching, and agriculture.

The biomass in the mature forest in Paragominas is approximately 350 t/ha, with some 300 tons of biomass above ground (Uhl et al. 1988) and 45 to 60 tons below ground (Nepstad 1989). We estimated that extensive-style logging reduces the total biomass pool by 29 t/ha, while intensive, management-based logging results in the emission of 67 t/ha of carbon (table 3.6). In comparison to logging, the conversion of forest to pasture results in an estimated biomass reduction of 335 t, which might liberate some 168 t/ha into the atmosphere. Average carbon losses associated with the conversion of forest for

TABLE 3.6 CARBON LOSSES ASSOCIATED WITH LOGGING, RANCHING, AND AGRICULTURE IN PARAGOMINAS

	Carbon Loss (tons/ha)
Extensive-style logging	29[a]
Intensive, management-based logging	67[b]
Ranching	168[c]
Agriculture	156[d]

[a]We consider that mature forest contains 350 tons of biomass (above and below ground; Nepstad 1989, Uhl et al. 1988). Extensive-style logging results in the destruction of approximately 25 percent of stand basal area (Veríssimo et al. 1992). This includes damage associated with logging roads but does not count what is harvested. Given the close agreement between basal area and biomass, we consider the loss of biomass to be 88 t/ha (0.25 × 350 t). Considering that 50 percent of this biomass is composed of carbon, some 44 t/ha of carbon might be released through decomposition processes in the years following logging. Over the 90-year recovery period, logged forests should reaccumulate this lost biomass. Hence, we consider that the average carbon deficit resulting from logging damage at any one time during the 90 years is approximately 22 t/ha. The average volume of wood harvested in extensive-style logging is 38 m^3/ha. At least 50 percent of this wood is transformed to saw dust and scrap and the carbon therein is liberated quickly. Considering that wood specific gravity averages 0.7, this amounts to approximately 6.65 t /ha of carbon released (38 m^3 × 0.5 × 0.7 × 0.5). Hence, total carbon liberation with this land use is approximately 29 t/ha (22 t + 6.65 t). This assumes that the carbon embodied in sawn wood products is gradually liberated over the 90-year production cycle at a rate roughly equal to its reaccumulation in the forest.

[b]Management-based logging results in a reduction of forest basal area and, we assume, biomass by 50 percent, to 175 t/ha (0.50 × 350). During the 30-year recovery cycle, we consider that the forest will reaccumulate 80 percent of its original biomass, reaching 280 t/ha. Hence, the average biomass present, considering the entire 30-year rotation, is estimated at 228 t/ha ([(280 − 175)/2] + 175 = 227.5). Biomass reduction, therefore, is 122.5 t/ha or 61 t of carbon. Loss of carbon in wood processing, following the rationale explained in note a, is estimated at 6.65 t. Hence, total carbon loss is estimated at 67 t/ha. To the extent that wood products might have a lifetime of greater than 30 years, our figure would overestimate carbon emission.

[c]Pastures in Paragominas contain about 15 t/ha of above- and below-ground biomass compared to 350 t/ha in the forests that they replace (Uhl, unpubl. data; Nepstad 1989). Hence, the carbon loss is estimated at 168 t/ha (335 t × 0.5). This assumes that all the carbon in the biomass decomposes (i.e., that none becomes incorporated as soil organic matter).

[d]There is generally a 10-year fallow between cropping episodes in slash-and-burn agriculture. Approximately 75 t/ha of total biomass accumulate during the fallow. At any one time, we consider that these farm-fallows have half the year-ten biomass (i.e., 37.5 t/ha). Therefore, the estimated carbon loss associated with converting a forest to a slash-and-burn agricultural field is 156 t [(350 t − 37.5 t) × 0.5]. We use the same value for permanent-style, perennial agriculture, since these systems frequently have about 40 t of biomass per hectare (Subler 1993).

slash-and-burn agriculture and intensive, perennial cropping are similar to ranching (table 3.6).

Among the proposals that have surfaced in recent years to reduce global carbon emissions, Schneider (1993) has recommended that the North consider paying Amazonian countries to preserve their forests because these forests serve as global carbon reservoirs. The carbon losses associated with Amazonian-style ranching (table 3.6) and the returns from these activities are frequently small (table 3.2). Meanwhile, estimates of the costs to reduce global carbon emissions range from $3.75 to $43.70/ton (Schneider 1993). Hence, northern, highly industrialized countries may find it more cost effective, in the short term, to pay the South to reduce deforestation than to undertake costly measures to reduce carbon emissions at home.

One approach would be to provide stipends to Amazonian governments that agree to halt deforestation by adopting sustainable forestry rather than replacing their forest with low-biomass-per-carbon pastures. The stipend paid by the North might be equivalent to the difference in the carbon stored in the biomass in postlogged forest compared to that stored after conversion to pasture. In the case of traditional, extensive-style logging, there are 146 t/ha of stored carbon in the biomass, compared to 7 tons in abandoned pastures (table 3.6, notes a and c). Hence, the difference is 139 t/ha. Considering that there are 1.6 million ha of forest in Paragominas, such a policy might require the payment of a stipend of $834 million, using Schneider's lowest estimate for the value of stored carbon (1.6 million ha × 139 t/ha × $3.75/t). If management-based logging were adopted, the policy would generate revenues of $600 million for the municipality (1.6 million ha × 100 t/ha × $3.75/ha).

There is presently no viable mechanism for executing this type of agreement in Paragominas Municipality or in the rest of the Amazon. Unfortunately, discussion over trades such as this one typically lead to cries from within Brazil for complete sovereignty in making land-use decisions. As this example shows, however, this kind of negotiation, if conducted properly, might result in significant economic and social advantages, as well as ecological gains, for both sides.

The release of carbon into the atmosphere is not the only environmental cost of deforestation. Altering the forest results in the loss of biodiversity, soil erosion, and increased fire risk. Designating forest areas for sustainable logging, instead of for other uses that result in more severe forms of forest-alteration, might address many of these problems as well.

Using Research Findings to Ask Questions Relevant to Land-Use Planning

The information that we have summarized thus far can be used at the scale of individual counties to address key planning issues in Amazonia and elsewhere in the developing world. To illustrate this point, we pose four basic questions and use the existing data from Paragominas to answer each one.

Q1: What Are the Total Annual Receipts from Land Uses in Paragominas Municipality?

We answer this question for the year 1992. Wood extraction and processing was the dominant economic activity in Paragominas at that time. There were 137 sawmills in the municipality, and each processed timber from, on average, 242 ha/yr. Hence, an estimated 33,000 ha were harvested, generating a total cash flow of some $92 million (33,000 ha × $2,772/ha; table 3.1).

Ranching and agriculture were practiced on some 352,000 ha (Watrin and Rocha 1992), with 97 percent of this area (341,500 ha) dedicated to ranching and the remainder (10,560 ha) actively farmed in 1992. If we assume that all ranching is extensive (true for all but a small fraction of the pasture land), the ranches of Paragominas generated annual gross returns of $10.6 million (341,500 ha × $31/ha; table 3.2), and returns from agriculture (almost all extensive slash-and-burn farming in Paragominas) were approximately $11.3 million (10,560 ha farmed × $1,079/ha; table 3.3). Hence, total gross returns for Paragominas Municipality from wood industries, ranching, and agriculture are conservatively estimated at $114 million for 1992. These activities directly employed some 10,000 people and resulted in annual profits of approximately $39 million (tables 3.1, 3.2, 3.3, and 3.5).

Q2: Are Land-Use Activities in Paragominas Municipality Sustainable?

An accurate profiling of present activities is the starting point for contemplating the future well-being of a region. With this information, one can ask if present-day land uses are sustainable. In the case of Paragominas, the answer is no. Profits from extensive-style ranching are low, and, in the absence of measures to rejuvenate pastures, ranches are eventually abandoned. A study of sixty Paragominas ranches revealed that 36 percent of the pasture area was degraded (EMBRAPA/CPATU 1989), and Watrin and Rocha (1992), working from LANDSAT images, reported that 43 percent of the deforested area of Paragominas was abandoned, apparently the result of nonsustainable ranching practices.

Extensive-style agriculture, likewise, is unsustainable unless long (ten-year) fallows are used. In many agricultural colonies in Paragominas Municipality, land is already in short supply and the fallow period is five years or less. Given such short fallows, agriculture yields gradually decline as soils lose their fertility (Sánchez 1976).

Wood extraction and processing, the main economic activity of Paragominas, is also on unsustainable footing. In 1992, the 137 sawmills in Paragominas required, jointly, 33,000 ha to supply their wood needs. Although there were some 1.6 million ha of virgin forest in the municipality in 1988, the actual area available for logging was closer to 1.4 million ha because an estimated 130,000 ha had been harvested from 1988 to 1992 and 60,000 ha were in Indian Reserves. Given present extraction rates, and assuming a closed system (i.e., that all mills harvest within the confines of the municipality), these wood stocks would be exhausted in thirty-four to forty years (1,400,000 ha/ 33,000 ha), long before the recommended ninety-year rotation cycle. This would not be the first time that wood industries exhausted forest resources in Amazonia. Three hundred miles to the South of Paragominas, the town of Xinguara is pocked with abandoned mills and the surrounding forest has been logged out. If this scenario is repeated in Paragominas, then the municipality's annual revenues would decline from $114 million (with logging) to $22 million (without logging) (figure 3.4). Of course, extensive-style ranching and farming could spread to the logged-out areas, temporarily increasing revenues, but eventually both the timber and soil nutrient resources will be mined to exhaustion, given present-day trends.

Q3: Are There Alternatives to the Present Development Model of Paragominas?

The data that we have compiled on the practices of timber industries, ranchers, and farmers can be used to consider alternative, municipal-level development scenarios. First, we consider scenarios in which Paragominas Municipality could become self-sufficient in terms of food supply (exercise 1), using either extensive and intensive methods of production. Then, we create simple scenarios showing how the intensification of each land-use activity could provide higher returns and more jobs for the municipality (exercise 2). Finally, we create a more realistic, diversified scenario (exercise 3) for implementing sustainable land-uses in Paragominas Municipality.

Exercise 1—Envisioning Self-sufficiency in Food Supply There are approximately 615,000 ha of deforested land in Paragominas. Approximately 55 percent of this area is occupied by extensive-style cattle

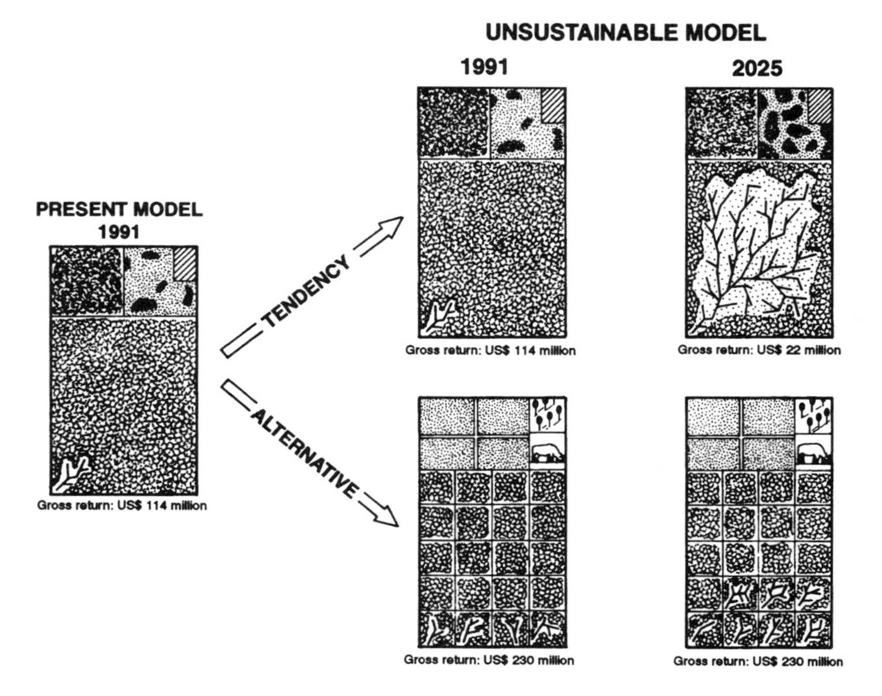

FIGURE 3.4 Comparison of the Current Extensive Land-Use Models in Paragominas and an Alternative, Intensive Model ***Note:*** Currently, the municipality realizes annual gross returns from logging, ranching, and agriculture (combined) of $114 million. However, if timber supplies become scarce, as current trends suggest, revenues might fall to $22 million/yr. The alternative model, which uses intensive land-uses, is projected to generate gross returns of $230 million/yr.

ranches, with almost all of the rest covered by secondary forest. Less than 2 percent of the deforested land in Paragominas is actively farmed at any one time.

In theory, the nutritional needs of the entire population of Paragominas could be met if land in the municipality that is already deforested (615,000 ha) were used for food production. It would take 668,000 ha to produce enough meat and starch (rice, beans, and cassava) to feed the municipality's population using extensive forms of production (table 3.7). In this case, 57 percent of the land would be

TABLE 3.7 AREA REQUIRED TO SUPPLY NUTRITIONAL REQUIREMENTS OF THE POPULATION OF PARAGOMINAS IN 1991, BY LAND-USE OPTION

				Extensive Model	Intensive Model
Diet Element	*Required Daily Intake (kcal) of Paragominas Pop.*	*Required Daily Food Consumption (kg) of Paragominas Pop.*	*Productivity of Ranching and Agric. (kg/ha) in Paragominas*[a]	*Area (ha) to supply population needs*	*Area (ha) to supply population needs*[b]
Meat	25,785,000	17,661[c]	17 and 57	379,192	113,092
Rice	62,984,000[d]	17,303[e]	830	91,310	7,609
Beans	62,984,000[d]	18,689	460	177,952	14,829
Cassava	62,984,000[d]	42,271	9,600	19,286	1,607
Total	214,737,000[f]	667,740	137,137		

[a]Annual live weight gain is 46 kg/ha in unimproved pasture (extensive model). Meat production is 36.6 percent of total production or 17 kg/ha. Annual live weight gain is 157 kg/ha (meat production = 57 kg/ha) in improved pastures (Mattos and Uhl 1994). Agricultural production data are from FIBGE (1985).

[b]To calculate the land area necessary to supply the population food needs, we divided daily consumption of each food type by production per hectare for that food type and multiplied by 365 days. For crops, in the extensive model, we multiply by 12 to allow for 11-year fallow between cropping episodes. For example, the land area necessary to satisfy the Paragominas population's rice needs following the extensive model is 91,310 ha [17,303 kg daily consumption/830 kg (production/ha) × 365 days × 12 yr.].

[c]To adequately meet the protein needs of the Paragominas population, we consider that that daily meat consumption should be approximately 190 g/person (i.e., 40 g of protein/person/day). This is eqivalent to 17,661 kg of meat [(92,355 people × 190 g)/1000 g]. This quantity of meat contains approximately 25,785,000 kcal (1 g of meat = 1.46 kcal; FIBGE 1981).

[d]We considered that the portion of the kcal supply not furnished by meat (i.e., 214,737,000 − 25,785,000 = 188,955,000 kcal) would be supplied by equal amounts of rice (3.64 kcal/g), beans (3.37 kcal/g), and cassava (1.49 kcal/g) (FIBGE 1981).

[e]Calculated as 62,984,000 kcal/3.64 kcal per g/1000 g = 17,303 kg; same calculation procedure used for beans and cassava.

[f]This estimate of daily kcal consumption for the population of Paragominas County considers population structure (sex and age groupings) and minimum daily protein requirements (Chaves 1985; Mitchell et al. 1978). In this example, the average daily intake is considered to be 2,325 kcal/person (214,737,000 kcal/92,355 people).

dedicated to raising cattle to supply meat. Using more intensive forms of land-use, however, the population's food needs could be met using only 137,000 ha, or 22 percent of the total deforested area (table 3.7).

The municipality's caloric and protein needs could be met even more easily through a diet that excluded meat. In this case, 328,000 ha (53 percent of the cleared land) would be required using extensive farming approaches and a mere 27,000 ha (4 percent of the deforested land) using intensive approaches (table 3.7). Although it may appear naive to contemplate a situation wherein an Amazonian municipality, such as Paragominas, would be totally self-sufficient in food production, there is an increasing demand for food in Amazonia as urban populations grow and it behooves planners to consider ways of promoting local production. As it stands now, most of the food consumed in the Eastern Amazonia comes from outside the region.

Exercise 2—Simplified Land-Use Models Recognizing that wood industries form the economic base of Paragominas Municipality, let us assume that this municipality decides to take measures to maintain the 1,400,000 ha in "working forest" (for logging and extraction of nontimber forest products) while setting aside 200,000 ha in Indian and biological reserves (about 10 percent of the municipality). This would leave available the 615,000 ha that have already been cleared for farming and ranching. With 1,400,000 ha of working forest, the annual timber harvest in the municipality would need to be reduced from 33,000 ha to 15,600 ha, assuming a ninety-year cutting cycle. In this case, annual gross returns would be $43 million; annual taxes, $5.6 million; and jobs, 2,640. However, if Paragominas' working forest estate of 1,400,000 ha were managed (thirty-year cutting cycle), it would be possible to harvest 46,700 ha/yr. Gross returns would increase to $129 million with concomitant increases in taxes to $15 million and jobs to 9,090, all the while implanting a sustainable timber resource-use approach (table 3.8).

Sustainable approaches to ranching also appear possible. If the entire deforested area of Paragominas Municipality (615,000 ha) were dedicated to extensive ranching, annual sustainable gross returns would be $19 million, tax generation would be $3 million, and some 6,500 jobs would be provided. If the same area were transformed to improved pastures, gross returns and taxes would jump to $64 million and $11 million, respectively (table 3.8).

On the other hand, if the entire deforested area of 615,000 ha were

TABLE 3.8 LAND-USE MODELS FOR PARAGOMINAS, SHOWING THE PROJECTED FINANCIAL RETURNS AND JOB GENERATION FOR PREVAILING AND ALTERNATIVE LAND USES[a]

	Sustainable Options					
	Wood Industries		*Ranching*		*Agriculture*	
	Prevailing Model	*Alternative*	*Prevailing Model*	*Alternative*	*Prevailing Model*	*Alternative*
	Extensive-style (90-year cycle)	*Management-based (30-year cycle)*	*Cattle Ranching (Unimproved Pasture)*	*Cattle Ranching (Improved Pasture)*	*Slash-and-Burn Agriculture*	*Perennial Cropping*
Total area available (ha)	1,400,000	1,400,000	615,000	615,000	615,000	615,000
Gross return ($/yr)	43,400,000	128,800,000	19,000,000	63,960,000	55,350,000	1,455,090,000
Profit ($/yr)	15,400,000	39,200,000	1,230,000	33,800,000	20,300,000	493,250,000
Start-up investment ($)	37,800,000	116,200,000	188,800,000	331,490,000	14,760,000	1,657,430,000
Tax generation ($/yr)	5,600,000	15,400,000	3,080,000	11,070,000	9,230,000	225,710,000
Job generation (*N*)	2,640	9,090	6,500	13,700	38,440	439,290

[a]Figures based on information presented in tables 3.1, 3.2, 3.3, and 3.5.

devoted to extensive-style, slash-and-burn farming, the sustainable return would be $55 million, with taxes projected at $9 million and job generation, 38,000. Finally, considering a complete conversion of this land to perennial cropping, the theoretical sustainable return would be approximately $1.5 billion; taxes, $226 million; and jobs, 439,000 (table 3.8).

These simplified models provide quick, order-of-magnitude comparisons among land-use options. In practice, individuals and communities, recognizing that extreme specialization renders them susceptible to unforeseen natural and market upsets, prefer to establish more diversified economies.

Exercise 3—More Realistic Models We can use the data to develop a somewhat more realistic and diversified model of sustainable development for Paragominas Municipality. We begin with the restriction that no more forest clearing be permitted in the municipality because logging is the economic engine of the region and cleared lands are already underutilized. Second, for the purposes of this example, we specify that beef ranching, insofar as it is generally extensive in style, be directed to properties larger than 100 ha, whereas slash-and-burn agriculture, insofar as it is generally practiced by small operators, be directed to properties of 100 ha or less. Given that there are some 60,000 ha in properties with less than 100 ha in Paragominas Municipality (FIBGE 1985), and that half of all properties must be maintained in forest under Brazilian law, we earmark 30,000 ha for extensive-style agriculture. The remaining 585,000 ha of cleared forest is designated for beef-cattle ranching in this example, with the 1,400,000 ha forest estate reserved for wood industries.

Given these land-use divisions and following extensive land-use approaches to logging, ranching, and agriculture, the municipality-wide estimated sustainable gross returns and profits are $64 million and $18 million, respectively, with 11,000 jobs generated. However, if forest management were practiced on the 1,400,000 ha of land reserved for forestry, improved ranching practices on the 585,000 ha of active and abandoned pastures, and perennial cropping and dairy farming, in equal portions on the 30,000 ha of agricultural land, gross returns and profits would jump to approximately $230 million and $84 million, respectively, and 25,000 jobs would be provided (figure 3.4).

Overall, the data available suggest that the economic and social situation of Paragominas Municipality could improve if more intensive and sustainable land-use approaches were adopted: gross returns were 2 times greater in the intensive/sustainable model outlined here compared to the actual situation; employment opportunities increased by 15,000, and there was an estimated $45 million increase in annual profits (figure 3.4).

We wish to stress, however, that the models we have presented are simplified versions of alternatives for this municipality's future. They treat Paragominas as a closed system, and they neglect any consideration of costs and benefits to the rest of the region and to the nation. Furthermore, the models are static. Downward or upward shifts in the sale price of beef, wood, or perennial crops and/or shifts in the cost of key inputs, such as fertilizer and machinery, would affect outcomes. Increases in land values could also cause shifts in land-use activities.

Many factors could and will surely change over the coming years. Clearly, such projections are not the end point of municipality planning in Amazonia, but they are a significant step in an important direction—namely, providing a quantitative basis for evaluating resource-use options and for making land-use decisions at the regional level, something heretofore little considered in Amazonia.

Q4: Where Might Capital be Found to Allow a Shift to Intensive, More Sustainable Approaches to Forestry, Ranching, and Agriculture?

Although alternative, more intensive approaches to forestry, ranching, and agriculture could mean significant increases in gross returns, profits, taxes, and jobs per hectare, capital is necessary for a switch to these more intensive practices (tables 1, 2, and 3). One measure of the feasibility of intensification is the number of years an activity needs to be practiced to pay off the cost of the start-up investment (figure 3.5). In the case of extensive-style wood industries, the annual profit was $961/ha, which would pay off the $2,391/ha initial investment in slightly more than two years. The pay-back period for management-based wood industries was about three years. Slash-and-burn agriculture paid off its initial investment in less than one year, while perennial-based agriculture would take three years (considering average returns over a ten-year period; table 3.3). By contrast, investors in extensive and intensive ranching would not recuperate the initial investment for fifty and ten years, respectively. Ranching, particularly in an extensive format, compares poorly with other options because of low productivity and high initial investments. Nonetheless, ranching is an effective way to lay claim to land

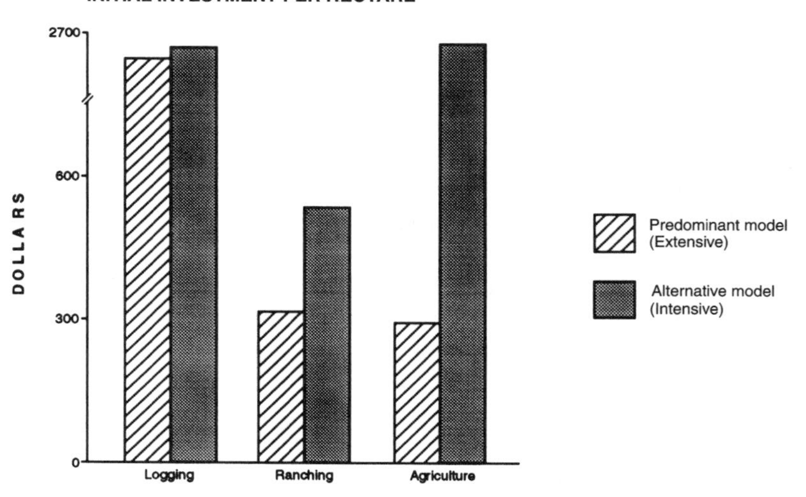

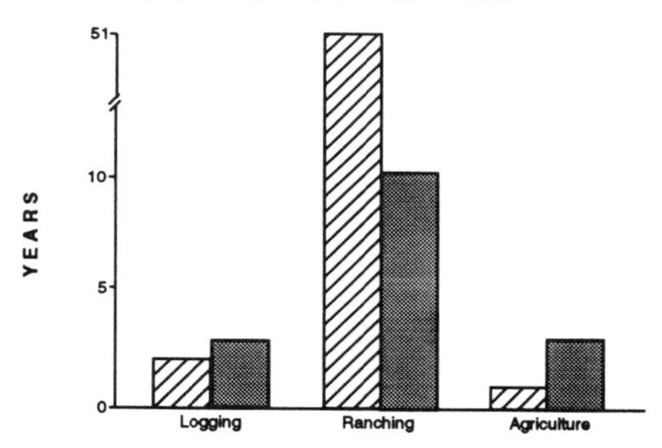

FIGURE 3.5 **Top:** Capital Investment Necessary to Start Operations in the Timber, Ranching, and Agricultural Sectors
Bottom: Years Required to Pay Back Initial Investments

(Hecht et al. 1988) and makes economic sense when one considers that land holdings in eastern Amazonia, even those that contain only degraded pastures, have been rapidly increasing in value in recent years as infrastructure develops.

CAPITAL SOURCES FOR INTENSIFICATION OF FORESTRY The prevalent, extensive-style logging that is practiced in Paragominas Municipality could be intensified by pursuing management-based approaches to forestry. This would entail an investment of approximately $113/ha in planning and silvicultural operations, and a waiting period of thirty years to receive the benefits (Barreto et al. 1993). In reality, wood industries can acquire logging rights from ranchers for $70/ha. In this case, they pay less than half the amount required for forest management and can extract timber immediately. But Veríssimo et al. (1992) have shown that, even if mill operators were obligated by law to carry out management practices, wood industry profit margins would remain above 20 percent, provided certain price guarantees were adhered to (see table 3.4). Nevertheless, given the abundance of standing timber, it is unlikely that logging companies will opt for management unless coerced by the government to do so.

Recognizing that forest management is not an attractive investment for wood industries, Paragominas Municipality could adopt a law to promote sustainable forestry. For example, the municipality might limit the amount of forest each mill could harvest in a given year. In this case, the maximum total annual harvest without management might be set at 15,600 ha (1,400,000 ha/90 yr). Given that Paragominas Municipality already has 137 mills, the allowable annual harvest per mill (assuming a closed system) would then be set at 114 ha (15,600 ha/137 mills), a value 47 percent of the present average harvest per mill. It might be further stipulated that those mills that strictly adhere to proper forest management practices be permitted to harvest 3 times this minimum area (340 ha/yr), since harvest intervals should be reduced from ninety to thirty years under management. Such a law would allow Paragominas to favor mills that practice forest management while penalizing those that do not, and, in so doing, it would help ensure that wood industries continue to have an important role in the economy.

CAPITAL SOURCES FOR THE INTENSIFICATION OF RANCHING As pastures age, forage grasses lose their vigor. Some ranchers in Paragominas Municipality are seeking ways of rejuvenating abandoned pastures, but the restoration process is expensive (table 3.2). Given that profits from extensive-style ranching are low (table 3.2), most ranchers, apparently, do not have the capital to finance pasture improvement.

Mattos and Uhl (1994) reported that ranchers in Paragominas are using wood sales in ranch tracts that are still forested to finance the rejuvenation of pastures. When ranchers sell logging rights, 3 to 4 ha are logged to restore 1 ha of abandoned pasture. When ranchers conduct their own logging, almost 1 ha of degraded pasture can be restored with the profits from each hectare of forest that is logged. Hence, ranchers do have an opportunity to intensify their operations insofar as they usually own large tracts of forest. The forest serves a vital function in the intensification of ranching by providing large chunks of capital from time to time to cover investments in infrastructure, animals, and forage improvements.

When forest timber is not available to finance intensification of ranching, it might be possible to plant annual crops together with pasture grasses on degraded lands that are being rejuvenated through tillage and fertilization. These annual crops are harvested before competition with grasses becomes severe. Veiga (1986) estimates that the yields from these crop plantings could cover the costs of pasture rejuvenation.

CAPITAL SOURCES FOR THE INTENSIFICATION OF AGRICULTURE The amount of land dedicated to agriculture in Paragominas Municipality is relatively small in comparison to that devoted to timber extraction and ranching. Most farmers practice slash-and-burn agriculture and have small lots (<100 ha), almost no access to capital, and low profits when compared to ranchers and wood industries. Moreover, farmers frequently live in small communities far from urban centers. The rustic roads linking these colonies to the Paragominas market are often passable only during the dry season. Provisions for health care and schooling are marginal. Generally, the only tools available to the farmer wishing to intensify are his or her labor (Toniolo and Uhl 1995).

As is the case with ranchers, selling logging rights to loggers can provide capital for farmers, but on a much smaller scale. Also, because farmers usually manage to find land only in remote locations and because they lack sophistication, they usually receive much less than the going rate for the timber on their small forest holdings.

Unlike wood industries and ranchers who have internal sources of capital for intensification, farmers, then, need access to outside sources of capital to make the shift from extensive slash-and-burn agriculture to both more intensive and sustainable alternatives. There are three sources of help. First, there is federal credit. The Constitutional Fund for the North (Fundo Constitucional do Norte) created in 1988 offers attractive credit options with a three-year payback period at low interest. However, the colonist must have title to his land to qualify, and most farmers do not. Furthermore, most small producers are unaware of the program. These problems could be rectified.[2] Second, municipal governments might help small producers.

Local governments can aid farmers by providing (1) seedlings of perennials, (2) short-term loans to cover the establishment costs of perennial systems, (3) vehicles to transport produce to markets at cost, (4) marketing facilities, and (5) guaranteed minimum prices in local markets. Finally, agricultural intensification might be stimulated through international funding agencies. Recently, there has been a flowering of grassroots organizations and peoples' associations in Amazonia. Foundations and funding agencies, such as Worldwide Fund for Nature, Ford Foundation, U.S. Agency for International Development, and the British Overseas Development Administration (now DFID) are helping these groups. Community projects aimed at sustainable farming using mixed perennial systems are viewed with particular favor by these international groups.

A LEADERSHIP ROLE FOR LOCAL GOVERNMENTS IN THE RESPONSIBLE DEVELOPMENT OF THE EASTERN AMAZON

There is heightened awareness and interest in environmental protection and resource-use planning in Brazil, especially at the federal and state levels. For example, federal legislation now requires that property owners who wish to deforest their land first get a license to do so, and, in the case of large land parcels, a complete environmental impact assessment must be conducted and approved before permission to deforest is granted. Meanwhile, Amazonian states, such as Pará, are now engaged in land-use zoning programs that aim to control and even prohibit the practice of certain activities (e.g., ranching) where they seem ill advised based on environmental criteria.

Municipal governments also might play a big role in environmental regulation and resource-use planning. Brazil's new federal Constitution, adopted in 1988, facilitates local land-use planning by providing greater administrative, financial, and legislative autonomy to municipalities (Gonçalves 1991). In addition to being granted control over city planning, municipal governments may now share opportunities with the state and federal government to provide health care, protect the environment, provide incentives for sound agricultural practices, regulate land settlement, and assess fines that discourage abusive practices (Aguiar 1991).

The First Step: Elaboration of a "Municipality Development Document"

At present, Paragominas and other Amazonian municipalities exert no authority over land-use activities within their borders. Farmers and ranchers wishing to clear forest and replace it with pastures or crops seek permission from the state-level Secretary of the Environment. Meanwhile, loggers wishing to extract timber from forest tracts have their plans approved by the Federal Institute of the Environment (IBAMA). However, there is now a provision in the National Environmental Legislation that allows municipalities to exercise control over land use, provided they formulate and adopt a "municipal planning document" (Sistema Nacional de Meio Ambiente, Decree 99.274, Article 19, 1990).

The elaboration of this municipal planning document entails the careful assessment of land uses (using remote sensing techniques), infrastructure (roads, schools, hospitals), services (water, energy), resources (locations of minerals, forests, waterways), and land-use potential (based on soils, relief, infrastructure, markets, and climate). At its heart, this plan is intended to be a zoning document, specifying and restricting how municipal lands can be used and providing both incentives and prohibitions to direct development, toward desired ends. The federal law specifies that this plan must be produced by a competent entity, such as the Municipal Council of the Environment. Once approved by the Municipal Assembly, the municipal planning document becomes law.

In the case of Paragominas, such a document could specify that all ranching and agricultural activities be restricted to areas that have already been cleared. Similarly, recognizing the importance of forestry activities to the local economy, the document could restrict the amount of the annual wood harvest to sustainable levels, specify management practices, and enforce these regulations.

Good Intentions Are Not Enough: The Need for Serious Enforcement

Simply having a development plan is not enough; simple, cost-effective enforcement techniques are also essential. The availability of low-cost satellite images represents a major breakthrough in land-use monitoring. These images allow the determination of location and extent of virgin forest, logged forest, and second-growth forest lands, and they allow the distinction to be made between extensive and intensive approaches to agriculture and ranching. And the cost is not great: complete LANDSAT coverage (1:250,000) of Paragominas Municipality (24,000 km$^{2)}$ is approximately $3,000.

To effectively monitor land use, property lines must be superimposed on these LANDSAT images. A study by M. Uzeda and C. Uhl (IMAZON, internal doc.) shows how this can be done using data available from INCRA (Federal Land Tenure Institute) and from local deeds offices. By crossing the land-use information (LANDSAT) with data on property limits (INCRA), municipalities can create a data base revealing, for each property holder, the amount of land in different land-use categories. Such a data base would allow the government to monitor how each landholder is using his property. For example, a landholder with two-thirds of his land in abandoned pasture, and the rest in logged-over forest, might be denied license to clear more land and fined because he violated the law that specifies that 50 percent of each holding must be maintained as forest. Likewise, a property holder who maintains a large percentage of his land in virgin forest might pay reduced land taxes or no tax at all on his forest land because the maintenance of virgin forest provides social benefits in the form of biodiversity preservation, carbon stocking, and barriers to fire spread. Although land taxes are under federal jurisdiction, counties might be able to make binding provisions for tax exemptions and/or tax increases in their municipal planning document.

The investment that Paragominas would have to make to develop a strong information base to monitor land-use and guide development decisions would be small (approximately $20,000 to establish the computer system and $50,000 to run it each year). With a team of four people (one researcher, one computer operator, and two inspectors) Paragominas could (1) operate a Geographic Information System (GIS) for land and resource-use monitoring, (2) evaluate and monitor ecological, economic, and social impacts of major land-use activities, and (3) conduct field checks to verify that resources are being carefully managed in conformance with regulations. Finally, an efficient system for monitoring land uses might also enable negotiations with developed countries concerning monetary compensation for the retention of forest as carbon reservoirs.

CONCLUSION

In its broadest expression, this chapter is about land-use planning in the Eastern Amazon, a region that has been haphazardly colonized over the last thirty years. We have shown that it is possible to compare the prevailing extensive-style land uses of the Eastern Amazon in economic, social, and environmental terms and to subject more intensive land-use alternatives to a similar scrutiny. When this information is analyzed at the scale of the municipality, citizens can assess the sustainability of current activities and evaluate alternative development options.

While it is heartening to observe that the information and tools to conduct land-use planning in Amazonia are becoming available, two grand impediments to regional land-use planning remain. The first is the lack of professionals trained in environmental science, resource economics, land-use law, and policy who are able to assume leadership roles in land-use planning at state and local levels. Fortunately, some universities are developing graduate programs to provide training in these planning disciplines, but much more needs to be done.

Even with a cadre of planning professionals, a second impediment looms large—the absence of a well-informed citizenry who could participate fully, through democratic processes, in the development, adoption, and implementation of rational land-use objectives. Fortunately, we note some stirrings on this front in recent years: Amazonian citizens are slowly abandoning passive ways for more active roles in community and regional decision making. For example, the formation of hundreds of grassroots organizations throughout rural Amazonia in the 1980s and 1990s shows that some rural people are working to gain a voice in political affairs. Natural and social scientists can help promote wise resource uses in Amazonia by providing holistic, objective analyses on important resource-use questions and by expressing their findings in didactic formats accessible both to policy makers and citizens.

NOTES

1. By the time this book reached publication, Paragominas had been subdivided into three municipalities, but this subdivision in no way affects our analysis, nor the conclusions therein. This is because we use Paragominas as a tool throughout this presentation to illustrate what might be done at the municipal level to promote resource stewardship in the Brazilian Amazon.
2. Indeed, as of 1995, the Constitutional Fund for the North was providing credit to farmers without official land title, provided that they were members in good standing of rural agricultural associations.

4

Charcoal-Fueled Steel Production in Brazil: An Exercise in Environmental Valuation

JOSEMAR X. DE MEDEIROS

One of the important debates in Brazil since the beginning of industrialization has concerned the prospect of developing major steel works as a motor to economic growth. This discussion began in the 1920s as a result of investment interest from the international steel industry. It was during these discussions that the concept of using charcoal rather than coal as a source of energy in steel smelting first gained currency, given Brazil's significant forest resource base and lack of coking coal deposits beyond those found in Santa Catarina. However, at that time, analysts rejected establishment of a major charcoal-fueled steel industry in Brazil. This was done for two reasons: First, no large-scale steel operation based on charcoal had previously been established in the industrialized countries and, second, its environmental effects were already considered a potential problem (Magalhães Gomes 1983).

Despite these concerns, the Companhia Belgo-Mineira was founded at the outset of the 1930s, becoming the first Brazilian integrated steel mill and the largest such mill fueled by charcoal in the world. Then, as well as today, the viability of charcoal-fueled steel production is considered to have been a result of the historically low wages prevalent in Brazil, which cheapened energy costs to the industry.

In the past two decades, because of the development of new reforestation technologies and, particularly, the introduction of fast-growing eucalyptus species in fuelwood plantations, the potential for expanded charcoal production has begun to be seriously contemplated. Many steel mills in Brazil now employ charcoal as a fuel and ore reducer. The most important such mills today are those of

Pains, Mannesman, COSIGUA, and ACESITA, each producing over 500,000 tons (t) of unprocessed steel annually. If the output of these mills is added to that of the numerous independent producers active in this sector, charcoal fuel contributes today to 30 percent of the national production of pig-iron, the first step in the processing of iron ore for steel manufacture.

The Brazilian steel industry has evolved through product diversification, ranging from ingots and semifinished materials to ironwork and flat, long and drawn shaped products. To simplify the analysis of economic and ecological aspects of this sector, this chapter focuses attention on pig-iron, in whose production charcoal represents a principal cost component and whose production implies significant environmental impacts.

ECONOMIC AND SOCIAL ASPECTS

In 1995, the charcoal-fueled pig-iron segment of Brazil reported gross output of US$4.2 billion, of which $1.6 billion was in foreign exchange credits. Around $660 million was attributed by this segment to its costs in charcoal production, marketing and transport. Overall, pig-iron production generated $546 million in tax revenues and created 158,000 direct and indirect jobs, of which 110,000 were in the production and transport of charcoal alone (ABRACAVE 1996).

Charcoal plays an important role in the national energy consumption structure; in recent years, in energy equivalent terms, its contribution has been maintained at a level similar to that of fuel-alcohol and gasoline, or on the order of 6 million t of petroleum-equivalent each year. Of the 10 million t of charcoal produced annually in Brazil, almost 70 percent is allotted to steel and pig-iron manufacture, to produce 7 million t of raw steel. The average price paid to producers for charcoal consumed in steel manufacturing varies from $12 to $20/m^3, depending on seasonality and the region of the country from which it is derived. About 34 percent of charcoal output goes to integrated steel mills and 66 percent to independent producers. Nearly all of Brazil's production of steel-alloys is based on charcoal.

Initially, charcoal production was concentrated in pig-iron producing centers in the iron-ore-rich state of Minas Gerais. As years went by, charcoal activities expanded, and today they are mainly concentrated in northern parts of the state of Minas Gerais, as well as in the so-called Triângulo Mineiro, in the western part of this state. It has also reached other states such as the south of Bahia in Brazil's Northeast, and the eastern part of Goiás and Mato Grosso do Sul in the Center-West region. In the 1980s, a new pig-iron pole began to develop along the Carajás railroad, in the northeastern Amazon region, where some eight pig-iron projects were installed and a number of additional mills approved. These benefited from development by Companhia Vale do Rio Doce of the enormous nearly pure iron ore deposits at the Serra de Carajás in southern Pará. At present, steel production in this region has reached around 1 million t/yr of pig-iron, all produced using charcoal derived from rapidly dwindling native tropical forests (including those in Paragominas; Chap. 3).

In general terms, charcoal derived from native forests has two economic motivations:

1. As an activity arising from or complementary to the expansion of the farming and cattle-raising frontier, it offers income to offset the cost of land preparation for establishment of crops or pastures.
2. As an autonomous economic activity, it generates a significant income to producers, and it is a source of employment, particularly in the dry season.

The first case is common in regions having economic potential for agriculture and cattle raising, such as the Triângulo Mineiro. Under these circumstances, if one compares forest management for charcoal with agricultural crop production as a land use, the latter is much more financially viable. Frequently, wood cut during land clearing is totally lost through burning, as a consequence of the landowner's hurry to introduce agricultural activities (Fundação João Pinheiro 1988).

The second case occurs in regions with lower economic potential, where charcoal production activities arose as a significant economic alternative. This was the case in northwest Minas Gerais, where the city of João Pinheiro became a great charcoal-producing center in the 1970s. Originally based on native forest exploitation, this activity is now supplied by fuelwood from tree plantations. In this region, reforestation for energy purposes has quickly developed at a large scale. Reforestation in such areas has been stimulated by low land prices, tax incentives, and the extremely low cost of wage labor. Today, a considerable part of João Pinheiro's hinterland is reforested with eucalyptus. Charcoal production in native forests as well as in reforested areas plays a substantial role in employment and income generation in this region. Charcoal plays a similar role in northern Minas Gerais, southern Bahia, and both Goiás and Mato Grosso do Sul.

ENVIRONMENTAL ASPECTS

The production of charcoal and its use in steel production has always been associated with environment devastation. This devastation arises from impacts observed during all stages of charcoal fabrication. The environmental effects are primarily related to deforestation of native species, establishment of homogeneous exotic tree plantations, and the transportation and use of charcoal. These activities lead to effects of a local and even global dimension. Soil, air, water, fauna, flora, and human societies are all significantly affected.

Today, about 654,000 hectares (ha) of native forests are deforested annually for charcoal fabrication (Medeiros 1995). This process has provoked severe pressures on ecosystems such as the Brazilian savanna (*cerrado*), where many animal and plant species have become extinct. Other consequences of this practice include gaseous emissions (CO, CO_2, particulates, etc.) from smoke due to forest fires set by farmers to clear their land, and increased soil erosion, leading to gullying and to alteration in the hydrologic regime.

Eucalyptus plantations destined for steel manufacturing now cover 2.4 million ha (Siqueira and Pierin 1990). The *Eucalyptus* genus contains more than 600 species, but only about a dozen of them have been used at plantation scale. Eucalyptus plantations are widely considered to provoke serious effects on the environment. Over the past thirty years, a number of authors have sought to identify and quantify the impacts of such plantations, principally in Minas Gerais (Lima 1993). There, problems associated with industrial forest plantations led the state government to intervene, enacting laws that limit the share of any municipality's territory on which homogeneous plantations may be established.

Charcoal production has commonly involved inhumane labor and adverse health conditions, exhausting work, and exploitation of child labor. Both the charcoal kilns themselves and the pig-iron blast furnaces, despite being less harmful to the environment than are coke-based plants, represent a source of pollutants such as CO, CO_2, particulates, and overthrow dust, and of solid residuals such as charcoal clinker and fines. Table 4.1 summarizes the environmental impacts

Table 4.1 Environmental Impacts of Production and Use of Charcoal

Activity	*Soil*	*Air*	*Water*	*Flora*	*Fauna*	*Humans*
Deforestation with Burning	Exposure with increased erosion; reduced organic matter	Increase in local temperature, particulates, and CO_2	Modification of river flow regime, runoff, and flooding	Reduction of biodiversity with species extinction	Habitat destruction, affecting animals, insects, and natural predators	Worsening of subsistence conditions of local populations
Homogeneous Plantations of Eucalyptus	Nutrient export, fertility reduction; compaction and allelopathic effects	Act as artificial windbreaks	River pollution due to percolation of chemical inputs and pesticides	Creation of artificial ecosystems inhospitable to native species	Creation of artificial ecosystems inhospitable to native species	Dislocation of food crops; dangerous working conditions

continues

TABLE 4.1 CONTINUED

Activity	*Soil*	*Air*	*Water*	*Flora*	*Fauna*	*Humans*
Charcoal Fabrication	Local soil degradation	Increase in CO_2, emission of condensible gases, heat, and particulate matter	Minor acid rain potential	Reduced potential photosynthesis at a local level	Creation of inhospitable local conditions	Unhealthy and dangerous work conditions
Handling of Fines and Charcoal	Local sterilization	Particulate emissions	River pollution	Reduction in potential photosynthesis due to dust deposition	Creation of inhospitable local conditions	Unhealthy and dangerous work conditions
Combustion of Charcoal in Blast Furnaces	Emission of ore waste and other residuals	Emission of CO_2 and other oxides, increased temperature and particulate matter	Minor acid rain potential	Reduction in potential photosynthesis due to dust deposition	Creation of inhospitable local conditions	Unhealthy and dangerous work conditions

TABLE 4.2 ENVIRONMENTAL IMPACT COSTS IN CHARCOAL-FUELED STEEL PRODUCTION

	Native forests		*Eucalyptus forests*	
Effect	*\$/ha*	*\$/t charcoal*	*\$/ha/yr*	*\$/t charcoal*
Soil erosion and nutrient losses	4.54	0.60	137.29	44.29
Labor displacement	5.63	0.73	—	—
Natural capital depletion	20.00	2.60	—	—
Increase of carbon dioxide	388.80	50.50	—	—
Water resource depletion	—	—	60.0	19.35
Water and atmospheric pollution	—	2.50	—	2.50
Total environmental costs	—	56.93	—	66.14

anticipated from the range of activities involved in charcoal-based pig-iron production.

AN EXERCISE IN ENVIRONMENTAL VALUATION

Contemporary discussion of environment valuation in the charcoal industry is closely related to efforts toward creation of forest user tax rates throughout Brazil. Beginning in 1994, the government of Minas Gerais, to reduce pressures on native forests as a source of charcoal, raised the tax levied on fuelwood, sharply differentiating between that originating from native forests and that originating from homogeneous tree plantations. This tax currently amounts to about \$1.70/m^3 for charcoal from native forests and around \$0.34/m^3 for plantations.

Discussion regarding the practical significance of these values would be enriched by quantitative estimates based on environmental valuation. With a view to valuating environmental impacts in the production and use of charcoal in steel mills, the analysis below selected six environmental pressure points listed among those in table 4.1 that have been quantified in physical terms through scientific studies carried out in various fields (silviculture, agronomy, economy, ecology, hydrology, etc.).

Table 4.2 provides a summary of the monetary valuations incorporated in the present study. These results suggest that the environmental costs involved in the different activities associated with production and use of charcoal in steel manufacture total \$56.93 and \$66.14 per ton of charcoal derived, respectively, from fuelwood originating in native or planted forests.

The following presentation describes the methodologies and valuation techniques employed for each of the environmental impacts selected for analysis. The exploratory nature of the present exercise must be emphasized. The study represents a search for methodological pathways that can lead to definition of some of the costs associated with environmental damage in this case. Although currently represented by the uncompromising terms *externalities* and *free assets,* such environmental costs may be due to society's rapidly approaching its physical limits to expansion and thus may threaten the continuity of some economic activities. Along the way, monetary values will be quantified with due consideration for the underlying values at stake.

However, before initiating discussion regarding the valuation of these environmental impacts, it is necessary to establish the physical parameters associated with forest production and productivity.

Several typologies of native forests and their capacity to produce wood must be considered to assess charcoal production from native forests. The characteristic vegetation formations found in charcoal-producing regions include several types of *cerrado,* denser gallery forests (*matas ciliares*), forested areas near the sea in remnants of the threatened Atlantic Forest (*Mata Atlântica*) and open or regenerating forests that are undergoing processes of secondary succession.

To express average productivity of charcoal originating from native forests, a weighted value is adopted, consistent with the amount of fuelwood derived from each distinct vegetation formation as well as with their relative contribution to total area deforested each year. This study employs a statistical estimate of deforested areas used in charcoal production (Medeiros 1995). Table 4.3 shows that the weighted average productivity of charcoal originating from native forests in Brazil is 7.7 t/ha (in volumetric terms, 30.8 m^3/ha). This is a one-time harvest; the management option, although conceivable, is rarely adopted in practice.

Productivity of homogeneous eucalyptus stands in Brazil varies according to edaphoclimatic conditions and, significantly, to how they are managed. A productivity of 25 steres (m^3) of cordwood/ha/yr (with three cutting cycles, each being a seven-year period with one cutting) can be considered a realistic average in Brazil. Under these conditions, average charcoal productivity from plantations is 3.1 t/ha/yr of charcoal (12.5 m^3/ha/yr).

The specific use of charcoal in Brazilian steel production may be

TABLE 4.3 RELATIVE OCCURRENCE OF THE DIFFERENT VEGETATION FORMATIONS USED IN CHARCOAL FABRICATION, AND WEIGHTED AVERAGE PRODUCTIVITY OF CHARCOAL FROM NATIVE FORESTS

Vegetation formation	*Share of deforested area in Brazil (%)*	*Productivity of charcoal (t/ha)*	*Participation in weighted average for 1 ha (t)*
Campo	19	43	8
Caatinga	3	9.0	3
Cerradinho	17	6.0	10
Mata	4	200	8
Cerradão	1	168	2
Cerrado	56	8.3	46
Weighted average per hectare native forest			7.7

From Medeiros (1995).

estimated as 0.875 ton of charcoal per ton of pig-iron (CEMIG 1988). Considering the average productivities of charcoal in native forests and in eucalyptus reforested areas, as previously mentioned, the relation between forest area and pig-iron production are shown in table 4.4. It is important to keep in mind the distinction between a one-time cut with land-use substitution in the case of native forest, and an at least conceptually sustained yield of eucalyptus based on a continuous rotation.

In the following sections, the available physical data regarding each of the six selected foci of environmental impact are evaluated to determine their significance and, through valuation, to assess their monetary implications.

Loss of Soil and Nutrients

The exposure of soil both due to deforestation and to land preparation for tree plantations is associated with major water and wind erosion impacts, leading to significant loss of nutrients and of the soil itself.

Deforested areas may be relatively more or less vulnerable to erosion depending on the use to which they are put. In whatever case, however, it may be considered that soil will be fully exposed to erosion in the first year following deforestation. In quantitative terms, soil and sediment losses can be defined according to distinct soil

TABLE 4.4 RELATION BETWEEN FOREST AREA AND PIG-IRON PRODUCTION

Forest Type	*Charcoal Productivity (t/ha/yr)*	*Pig-iron Production from 1 ha of Forest (t/ha)*	*Forest Area Necessary for 1 t of Pig-iron Production (ha/t)*
Native	7.7[a]	8.8	0.114
Eucalyptus	3.1[b]	3.5	0.286

[a]Native forest productivity is an unsustainable, one-time only extraction, accounted only for 1 year, on the event of deforestation.
[b]Sustainable productivity of the eucalyptus forest refers to the annual average for a cutting cycle of seven years.

types. Data regarding the magnitude of such losses are provided by Hernani (cited in Barros and Novais 1990).

The environmental impacts of soil and sediment losses caused by deforestation may be assessed with reference to the substitute land use. In the case of agriculture or ranching, in which charcoal is produced only during the first year—the deforestation phase—the environmental cost attributable to charcoal is thus assessed only in the first year. This approach is followed despite the obvious fact that, unless the forest is allowed to regenerate, the substitute land uses will continue to result in erosion. The volume of soil and sediment losses from removal of vegetation in native forests, such as the *cerrados,* are here approximated using the value, shown in table 4.5 for the clearing process based on burning common in these regions, of 234.8 kg/ha.

In establishing eucalyptus forests to produce charcoal, the soil and sediment losses in the first year, the phase of clearing and preparing the soil, may be likened to those shown in table 4.6. For subsequent years, during which the soil is gradually covered by eucalyptus, these losses will be significantly reduced, rising again in the harvest period. Considering exploitation of a eucalyptus forest under *cerrado* conditions, with three cuts (one in each of three seven-year cycles), the probable losses of soil and sediment in average terms will be of 8 t/ha/yr. Table 4.7 presents a summary of probable losses of soil and sediment arising from charcoal produced from both native and plantation forests.

Besides the environmental costs arising from soil depletion, a good part of the eroded soil will end up in rivers and reservoirs. Water serves a range of functions including household consumption and

TABLE 4.5 LOSSES DUE TO EROSION OF SOIL, NUTRIENTS, AND ORGANIC CARBON BY DIFFERENT METHODS OF SECONDARY FOREST CLEARING

Clearing Methods	*Runoff Volume (mm/yr)*	*Losses (kg/ha/yr)*					
		Soil	*P*	*K*	*Ca*	*Mg*	*Organic C*
Slashing	139.2	130.4	0.02	4.81	2.92	0.45	4.2
Burning	208.0	234.8	0.09	4.80	6.58	1.11	18.5
Destumping	307.9	1893.3	0.10	4.89	12.05	2.16	288.7
Regeneration	2236	52.2	0.02	7.94	4.66	0.84	0.5

From Hernani et al. (1987), cited in Barros and Novais (1990:273).

TABLE 4.6 LOSSES OF SOIL AND WATER FROM A YELLOW-RED PODZOL SANDY AND MEDIUM-TEXTURED SOIL, BY DIFFERENT LAND PREPARATION METHODS[a]

Methods of Preparation	*Losses*	
	Soil (t/ha/yr)	*Water (mm/yr)*
Five-toothed scarification	22.5	121.1
Three-toothed scarification	26.4	127.9
Light grading	25.4	127.3
Disc plowing	49.8	211.6
Heavy grading	56.2	222.6

From Castro et al. (1986), cited in Barros and Novais (1990:273).
[a]Rainfall during experiment: 1392.5 mm/yr.

TABLE 4.7 ESTIMATED LOSSES OF SOIL AND SEDIMENTS ARISING FROM CHARCOAL FABRICATION IN NATIVE AND PLANTATION FORESTS

Type of Forest	*Losses (kg/ha/yr)*					
	Soil	*P*	*K*	*Ca*	*Mg*	*Organic C*
Native	234.8	0.09	4.80	6.58	1.11	18.50
Plantation	8000.0	0.42	20.64	50.86	9.12	1218.32

From Barros and Novais (1990:273), adapted by author.

electrical generation. Environmental costs of sedimentation are thus associated with a decline in the stored volume of water and the consequent decrease in the generation of electrical energy, as well as in the useful life of hydroelectric facilities.

Lake and Shady (1993:9) estimate that the secondary costs of erosion are at least twice as large as those directly associated with losses in productive soils. (These authors calculate that the damage caused by erosion to reservoirs in the United States amounts to around $10 billion/yr.)

Valuation of Soil and Nutrient Losses

Based on the preceding information, we observe that the environmental impacts associated with increased erosion in native as well as in plantation forests can lead to a significant loss of soil nutrients and also to a reduction of stored water volume in dams located downstream. The soil nutrient losses can be valued from the available physical data by calculating their replacement cost based on market prices of fertilizers traditionally used in agriculture. This procedure makes it possible to calculate the environmental costs based on soil nutrient losses, both in native and reforested forests. Table 4.8 shows the results.

The valuation of reduced water storage capacity in hydroelectric reservoirs due to silting can be accomplished by calculating the reduction of generating potential of a typical hydroelectric unit up to its operating horizon.

For this estimation, a typical hydroelectric facility is considered, with installed power generation capacity of 1200 megawatts (MW), a reservoir with flooded area of 1200 km^2 and an average depth of 10 m for a total volume of 12 billion m^3, operating at one-third of the reservoir's initial capacity and built at a cost of $2.4 million for each kilowatt (kW) installed. It is possible to infer the environmental cost caused by the silting process caused by erosion in a specific unit of the hydrographic basin, based on the rate of siltation, having as an operating limit the point when two-thirds of the reservoir volume has been silted up.

In this way, the environmental cost is estimated by calculating the relationship between soil losses from 1 ha in such a hydrographic basin and the total that would cause silting up of reservoir capacity, and the share of this fraction in the total facility investment. The result is that, for each hectare of plantation forest, there will be a depreciation of $2.88/yr in the hydroelectric unit. With regard to silting caused by deforestation of native forests, the cost is estimated at $0.09/ha.

TABLE 4.8 COST OF NUTRIENTS LOST THROUGH EROSION IN DEFORESTED NATIVE FOREST AREAS AND EUCALYPTUS PLANTATIONS

Substrate or Nutrient	*Native Forest* (kg/ha)	*Native Forest* ($/ha)	*Homogeneous Plantation* (kg/ha/yr)	*Homogeneous Plantation* ($ha/yr)
Soil	234.80	n.a.[a]	8,000.00	n.a.[a]
Phosphorus (P)	0.09	0.11	0.42	0.50
Potassium (K)	4.80	2.11	20.64	9.08
Calcium (Ca+Mg)	7.69	0.38	59.98	3.00
Organic carbon (OC)	18.50	1.85	1218.32	121.83
Total	—	4.45	—	134.41

[a]Not evaluated.

The valuation of this environmental impact, as shown in table 4.2, is $4.54/ha or $0.60/t of charcoal originating from native forests and $137.29/ha/yr or $44.29/t of charcoal from reforestation. The significant difference between these values arises from the difference between land clearing techniques: Whereas the traditional slash-and-burn technology has a fairly low impact, the destumping and grading associated with plantation establishment imply a far more severe erosion rate.

Labor Displacement

The living conditions of the local population are worsened when native forests are destroyed. Diminution of extractive resources used for food and raw materials, and the lack of alternatives to absorb the surplus labor created in this process, contribute to rural exodus and the consequent formation of slums in major urban centers.

Generally, the natural diversity of flora and fauna in the *cerrado* represents an important source of food, including protein; collection of fruit and oilseeds and, particularly, subsistence hunting are common. These resources also represent an important source of raw materials such as vegetal fibers used in the manufacture of household utensils, building materials used in rustic habitations, natural colorants, and so on. In fact, in areas having low population density, such as the *cerrados,* these extractive activities represent an important source of subsistence, complementing local cash incomes.

On the other hand, the removal of the native vegetation cover, owing to the expansion of the agricultural frontier or to the establishment of major forest projects, has often not brought economic benefits of development to the local population. In reality, the expansionist economic model, based on large projects for intensive monocrop cultivation or ranching, follows the logic of income concentration and rarely employs the displaced wage labor.

> The development of capital in intensive agricultural areas was unable to generate enough employment in order to absorb the wage labor surplus, causing migratory waves to the cities. . . . The modernization caused notable changes in labor relations, leading to a partial and precarious employment of former sharecroppers, *colonos* and *moradores* who were transformed into temporary and migrant laborers by the expansion of monocrops and the consequent increase of seasonality in employment. The expanded monocrops suppressed subsistence activities by agricultural workers, turning them into a surplus labor force seeking jobs in rural areas or in cities.
>
> *Brasil (1991:35)*

In conclusion, it may be stated that deforestation made survival by local populations more precarious and that the economic activities that followed deforestation, based on modern agricultural and silvicultural production systems, cannot absorb the wage labor freed up in the process.

Valuing Local Wage Labor Displacement

According to Stout (1980:43), the geographical area necessary to maintain one person on an extractive basis in savanna regions or forest formations such as the *cerrado* would be around 150 ha. Considering that extractive activities serve as a complement to other economic activities in this region, such as agriculture, it is possible to assume that the native forest area necessary to complement other subsistence sources could be estimated as 75 ha/person. This implies that, for every 75 ha of forest removed, one person would be displaced from his natural habitat. If, through this process, one of every two displaced persons is incorporated in the ranks of employed workers, the result is that the displaced laborers would be transformed into potential migrants at a rate of one person for each 150 ha deforested.

If we consider that about 650,000 ha are utilized annually in Brazil for charcoal fabrication, such activities would be responsible for the displacement of around 4,300 people who would be destined for the periphery of large cities. It is furthermore considered that such migrants would not contribute to productive economic activities during the year in which they are displaced. If we take the social cost of each adult person as thirteen minimum wages per year, around $845.00, based on the common pay scale current in rural areas of Brazil, we verify that the socioenvironmental cost of this impact would be $5.63/ha or $0.73/t charcoal, as shown in table 4.2.

Natural Capital Depletion

The exploitation of native forests on a nonsustainable basis causes a "natural capital depletion" represented by the loss of the finite natural resource of native forests. The deforestation carried out for charcoal production or plantation establishment does not consider native cordwood as scarce and finite natural capital. In agricultural frontier regions, native forests are perceived as a difficulty to be overcome, and deforested lands receive higher market values than those covered with natural vegetation. Landowners often choose not to use the wood to fabricate charcoal, as they are anxious to prepare the soil to initiate agricultural and cattle raising activities. This situation can be observed in the Triângulo Mineiro region:

> In this region, charcoal production is almost always an attempt to make use of the trees pulled up by tractors and chains in order to establish pastures and crop land. However, this does not always happen. Because of the high productive capacity of the soil, profitable charcoal fabrication is not often possible; charcoal is slow, takes time and, for example, an area of 250 ha would take 14 or 15 months to be cleared. But people have urgency and although they program deforestation over the period of one year, after six months they have it all deforested. So, what is to be done? Set the wood on fire. And an enormous forest area is burned.
>
> *Fundação João Pinheiro (1989:29)*

More recently, in regions proximate to major charcoal consumption centers in the Belo Horizonte and Sete Lagoas regions of Minas Gerais, where native forests were almost totally extinguished, fuelwood has begun to acquire a more substantial market value. Although less concentrated and inferior to the steel sector in consumption volume, there is a strong fuelwood consuming market represented by small ceramics firms, bakeries, and grain-drying warehouses. Only in regions like these does native timber have some value, but even in these cases the stumpage price of the wood has only a small role in the formation of fuel prices in the consuming market. According to the ABRACAVE cost structure for the manufacture of charcoal originating from native forests, the stumpage value in October 1993 was estimated at $1.00/stere, for a production

TABLE 4.9 PRICES OF EUCALYPTUS FOREST PRODUCTS

Products	*Stumpage Value ($/m³ st)*	*Final Price ($/m³ st)*
Charcoal	5.00	21.43
Market eucalyptus	7.50	20.00
Sale for cellulose production	9.00	21.00
Export roundwood	6.50	45.00
Sawn wood	21.00[a]	150.00[a]

[a]US$/m³
st, steres.
From Magalhães (1993:248).

cost at the charcoal kiln of $12.03/m³ of charcoal. As a rule, little or no value is given to the wood arising from deforestation.

Valuing Natural Capital Depletion

To give a market value to the forest resources in question, the price of wood derived from eucalyptus plantations destined for charcoal production can be used as a base. Most of the forest area planted to eucalyptus in Minas Gerais was originally destined for charcoal-fueled pig-iron mills. Since then, there has been a tendency for the forest sector to search for multiple uses for dense forest clusters to ensure a greater return on investment. The constant expansion of the pulp and paper industry and the greater profitability of this sector led to its becoming, unquestionably, a potential competitor for eucalyptus wood in charcoal production (Magalhães 1993:248). Table 4.9 shows the prices attained by eucalyptus wood.

The cost of planting eucalyptus used in charcoal production is on the order of $5.00/stere. Next, a coefficient is derived relating eucalyptus volume to native fuelwood used in charcoal production: 1 m³ of eucalyptus wood is equivalent to 1.5 m³ of native wood, a result of the greater uniformity of the former source. Considering these parameters, the environmental cost of depleting the natural resource may be treated as $3.33 for each cubic meter of native timber, or $10.00 for each cubic meter of charcoal.

Irreversible damages to native forests may be considered to represent natural resource depreciation or "loss of sustainable income" (see chapter 2). In this case, considering the natural productivity of biomass in native forests such as the *cerrado,* for example, estimated at about 6.0 m³/ha/yr for a regenerative period of 20 years (Thibau 1972:17), an average income loss of around $20.00/ha/yr may be defined for deforested areas. (See environmental valuation data summary in table 4.2.)

Increase of Carbon Dioxide in the Atmosphere

The removal, burning, and conversion to charcoal of natural vegetation alters the carbon sequestered in biomass, which may increase net atmospheric CO_2 accumulation, thus contributing to the greenhouse effect and the rate of global climate change.

The replacement of natural ecosystems by crops, pastures, or planted forests implies variation in biomass. The reduction of the original biomass storage implies net emissions of carbon in the form of CO_2 to the atmosphere. Increase of biomass storage in a particular area leads to a sequestration of carbon from the atmosphere, captured in the form of biomass.

Should a reforested area be periodically cut for fuelwood, varying the biomass storage as time goes by, it may be conceived as acting as a dynamic carbon sink. In contrast, a native forest area may be conceived as a permanent carbon sink because, when it reaches its climax, biomass storage remains stable. From the biomass balance perspective, as time goes by, managing fuel forests under short cycle rotation approximates this activity to that of conventional agriculture, with a much smaller average resident biomass storage if compared to an adult forest at climax. At the initial planting phase, the system has no biomass storage. By the eve of cutting, after the necessary growing cycle, it has achieved maximal biomass buildup, and after cutting, it returns to the initial situation of practically no accumulated biomass storage, and then the cycle starts again. In this sense, average biomass storage can be calculated by dividing the biomass production accumulated during a full forest growth cycle, by the number of growing years in that cycle; the result is equivalent to the quantity of biomass attributable to the average annual growth.

The present exercise estimates the immobilized or liberated carbon, considering vegetation cover in each of the previously presented forest types. Weighted-average values are used to determine the proportion of these types found over all deforested areas in Brazil. The carbon balances resulting from the net of immobilized and liberated carbon due to removal of different natural vegetation cover types and their replacement by eucalyptus fuelwood plantations are presented in table 4.10.

There are two ways to compare reforestation for energy purposes in terms of CO_2 balance:

TABLE 4.10 CARBON BALANCE FROM EUCALYPTUS REFORESTATION AS COMPARED WITH ORIGINAL BIOMASS AND DIFFERENT NATIVE FOREST TYPOLOGIES

Vegetation Typologies	*Participation in Deforested Area (%)*	*Original Carbon Stock*[a] *(t/ha)*	*Carbon Immobilized in Eucalyptus Plantation*[b] *(t/ha)*	*Carbon Released (t/ha)*
Campo	19	12.1	7.4	4.7
Caatinga	3	26.6	7.4	19.2
Cerradinho	17	12.1	7.4	4.7
Mata	4	90.5	7.4	83.1
Cerradão	1	59.2	7.4	51.8
Cerrado	56	24.2	7.4	16.8
Weighted average	—	23.0	7.4	15.6

[a]Refers to native forests in climax.
[b]Refers to annual average stock during a cycle of 21 years, with cuts in the 8th, 15th, and 21st years. The following parameters were used: annual average increment of 25 m^3 st/ha; apparent density of fuelwood from *mata* (0.7 t/m^3), *cerradão* (0.55 t/m^3), and *cerrado* and other types (0.45 t/m^3), and from eucalyptus (0.55 t/m^3); fuelwood/aerial biomass coefficient (0.65).

1. Consider that the original biomass in the area to be reforested no longer exists. This would apply to areas already exploited for agriculture or ranching activities or to degraded sites. Currently, such situations are widespread. In this case, the formation of an energy forest, even of short cycles, would represent a net capture of atmospheric CO_2.
2. Consider that the original vegetation cover exists in the area where the energy forest is to be established (the situation just described conceptually). In this case, the creation of an energy forest would imply net liberation of CO_2 to the atmosphere.

Table 4.11 presents the CO_2 balance for the replacement of native forest by annually cultivated grains such as soybeans and corn, under Brazilian *cerrado* conditions. Table 4.12 presents the CO_2 balance for the replacement of native forests by pastures in the previously mentioned conditions.

To estimate the annual volume of carbon liberated due to charcoal production for the steel industry, the deforested area, calculated at around 650,000 ha/yr, is used as a base. Considering the objective

TABLE 4.11 CARBON BALANCE FOR AGRICULTURAL ACTIVITIES (CORN-SOYBEAN ROTATION) IN RELATION TO ORIGINAL BIOMASS AND DIFFERENT NATIVE FOREST TYPOLOGIES

Vegetation Typologies	*Participation in Deforested Area (%)*	*Original Carbon Stock*[a] *(t/ha)*	*Carbon Immobilized in Cropping System*[b] *(t/ha)*	*Carbon Released (t/ha)*
Campo	19	12.1	7.2	4.9
Caatinga	3	26.6	7.2	19.4
Cerradinho	17	12.1	7.2	4.9
Mata	4	90.5	7.2	83.3
Cerradão	1	59.2	7.2	52.0
Cerrado	56	24.2	7.2	17.0
Weighted average	—	23.0	7.2	15.8

[a]Refers to native forests in climax.
[b]Considers biomass produced in terms of dry matter (grain + crop residues), during an annual corn-soybean rotation. Average productivities for corn of 5,000 kg/ha and soybeans of 1,800 kg/ha were used. The correlations between grain and agricultural residues were based on average coefficients suggested in National Academy of Sciences (1977:39).

of fuelwood self-sufficiency by 1999 imposed on major consumers in Minas Gerais, as required by law 10561/91, it has been estimated that 200,000 ha/yr of eucalyptus plantations would be needed to provide sufficient charcoal to meet demands by this deadline. It thus seems reasonable to treat this as the deforested area annually destined for establishment of eucalyptus energy forests. The remaining deforested area, 450,000 ha/yr, is assumed to be destined in equal parts to the formation of pastures and to agricultural activities.

Besides this, about 250,000 ha/yr of the 2.4 million ha of existing eucalyptus forests is cut for charcoal destined for pig-iron mills. The quantity of carbon liberated by cutting each year will be absorbed or immobilized by the growth in forests in the remaining area.

Table 4.13 presents a consolidated annual carbon and CO_2 balance, showing the amount liberated to the atmosphere by deforestation and the consequent use of the soil by other economic activities.

Valuing the Increased Carbon Dioxide in the Atmosphere

Based on the quantities of carbon or CO_2 liberated by deforestation, by silvicultural, agricultural, and ranching activities, it is possible to attribute the cost to society of such land uses. To do so, it is necessary

TABLE 4.12 CARBON BALANCE FOR RANCHING ON PLANTED PASTURE, IN RELATION TO ORIGINAL BIOMASS AND DIFFERENT NATIVE FOREST TYPOLOGIES

Vegetation Typologies	*Participation in Deforested Area (%)*	*Original Carbon Stock*[a] *(t/ha)*	*Carbon Immobilized in Planted Pastures*[b] *(t/ha)*	*Carbon Released (t/ha)*
Campo	19	12.1	6.1	6.0
Caatinga	3	26.6	6.1	20.5
Cerradinho	17	12.1	6.1	6.0
Mata	4	90.5	6.1	84.4
Cerradão	1	59.2	6.1	53.1
Cerrado	56	24.2	6.1	18.1
Weighted average	—	23.0	6.1	16.9

[a]Refers to native forests at climax.
[b]Considers carbon contained in dry matter, referring to an average, well-managed pasture using the principal grass species planted in the region (Pupo 1981:20).

to know the environmental costs associated with atmospheric emissions of carbon or CO_2.

> Among many recent proposals to reduce carbon gas emissions to the atmosphere, one is to use forests as carbon sinks, and to avert deforestation. Schneider [1993], evaluating the returns per hectare of Amazonian agriculture and comparing it to the cost of reducing carbon gas emissions in countries of the northern hemisphere, shows that the exchange is profitable for both sides at values lying between $3.75 and $43.70 per ton of carbon released.
>
> *Almeida and Uhl (chapter 3).*

Silvicultural, agricultural, and ranching activities in the *cerrado* region may be more profitable than those reviewed by Almeida and Uhl (chapter 3) for the eastern Amazon. It may also be more expensive to avert degradation of native forests in that region. Nevertheless, a conservative mean value of $24.00 for the cost of 1 t of carbon liberated to the atmosphere may be adopted based on the previously described criterion. The weighted-average values in table 4.13 lead to an estimate, therefore, of an environmental cost due to increased atmospheric CO_2 arising from deforestation, of $388.80/ha. These values are attributable only to lands deforested for agriculture, as the net carbon balance of eucalyptus plantation establishment is estimated

TABLE 4.13 QUANTITY OF CARBON AND CO_2 RELEASED ANNUALLY TO THE ATMOSPHERE, DUE TO SUBSTITUTION OF NATIVE FORESTS BY AGROSILVOPASTORAL ACTIVITIES FOLLOWING CHARCOAL EXPLOITATION

Substitute Land Use	*Area (1000 ha/yr)*	*Carbon Released to the Atmosphere*	
		C (t/ha)	*CO_2 (t/ha)*[a]
Agriculture	225	15.8	57.9
Eucalyptus	200	15.6	57.2
Pastures	225	16.9	62.0
Weighted average[b]	—	16.2	59.4

[a]The CO_2 molecule contains 27.27 percent carbon by weight.
[b]Weighted average in relation to areas destined for each land use, for a total of 650,000 ha deforested annually in Brazil.

to be approximately nil. That is, even if we assume that a portion of the area dedicated to eucalyptus plantations had to be cleared from native forest, the weighted emissions after accounting for carbon sequestered in the planted forest is fairly close to zero.

Water Resource Depletion

Eucalyptus forests can significantly reduce the water availability from hydrographic basins, particularly in those areas of sparse vegetation cover such as the *cerrados* and grasslands.

Among all environmental impacts related to large-scale eucalyptus plantations, those associated with rainfall, soil, and subsurface water have been generally recognized even by the defenders of this important tree species. The effect on rainfall occurs mainly as a result of interception by the tree canopy.

> A more significant hydrologic effect associated with precipitation is related to the process of rain interception by which the incident precipitation is redistributed by the tree canopy and is partially lost by direct evaporation of the intercepted water. . . . In reality, comparing the effects of reforestation of areas of open fields, pastures, or any other open vegetation areas, the estimated reduction found in the water productivity of the hydrographic basin is of approximately 20 percent, while the evaporative losses of the basin might, eventually, almost double.
>
> *Lima (1993:54)*

Even considering that the rain interception effect may also be manifested with similar intensity in other species used in reforestation and also in denser native forests, it is already known that, when compared with typical vegetation formations such as *cerrado, cerradinho,* and *campo limpo,* this effect implies significant loss of water productivity at the hydrographic basin level.

Beyond the interception of rainfall, the effect of eucalyptus on soil and subsurface water has been one of the most frequently discussed issues associated with the environmental impacts of this species. The available bibliography on this issue is vast and controversial. The view predominates that eucalyptus is capable of absorbing subsurface water more intensively than other forest species. In an experiment carried out in Minas Gerais, Lima (1993:85) reports a reduction of around 230 mm in drainage from a native *cerrado* area compared to the same area of a five-year-old *Eucalyptus grandis* plantation. In spite of the author's suggestion that this reduction is offset by the greater rate of biomass production in eucalyptus, the fact is that water productivity in the hydrographic basin will be commensurately reduced for this reason. Although such losses would be smaller if eucalyptus areas were to be compared with the denser vegetation found in forests and *cerradões,* the *cerrados* and other sparsely vegetated formations represent the principal share of deforested area used for charcoal production (Medeiros 1993:112).

Considering the annual average precipitation in the *cerrado* region of about 1200 mm, and taking into account the previous observations, an average estimate of water loss would suggest a reduction in water productivity of a hydrographic basin located in these regions of about 300 mm or 3.000 m^3/ha that would be deducted from that available to streamflow and downstream reservoirs.

Environmental Valuation of Water Resource Depletion

The reduction of water productivity in a hydrographic basin will have negative repercussions on uses such as household consumption and irrigation and, on a larger scale, electrical generation in power stations located downstream from the hydrographic basin in question. The negative effect on electrical generation will be used as a parameter to estimate the social cost associated with this environmental impact.

If it is considered that, of this reduced water volume, around 60 percent would be used to generate electricity in a given year, the reduction of water volume in each hydroelectric unit located downstream from the hydrographic basin would be of 1800 m^3/ha/yr of area planted to eucalyptus. This volume of water, in a representative hydroelectric power station of 60 m height, with 85 percent of genera-

TABLE 4.14 INPUTS AND EMISSIONS FROM PRODUCTION OF CHARCOAL-FUELED PIG-IRON

	Quantity (kg/t pig-iron)
Inputs	
Charcoal	875
Iron ore	1,600
Limestone	100
Quartz	65
Manganese	40
Air[a]	4,300
Emissions	
Clinker	150
Fines	40
Blast furnace gas	1,730 (excess)
Gas removal	4,060 (used in heating)

From CEMIG, 1988:150.
[a]Used in blast furnace combustion and heating.

tion efficiency, would produce around 255 kilowatt-hour (kWh) of electricity per year. If we consider at least four hydroelectric power stations operating downstream of any given plantation area, the losses of electric energy from the hydrographic basin would be around 1.0 MWh/yr.

To value this energy loss, we apply the marginal cost of electricity generation provided by the electricity sector, which is currently \$60.00/MWh. On this basis, the environmental cost of the water resource depletion due to reforestation with eucalyptus may be estimated at around \$60.00/ha/yr, as shown in table 4.2.

Water and Atmospheric Pollution

The use of charcoal in steel mills causes the generation and emission of pollutants such as CO_2, CO, particulates, dust, and solid depositions such as charcoal clinker and fines.

In terms of consumption, it was found that for each ton of pig-iron produced, charcoal-fueled steel mills require the raw material inputs and generate the emissions described in table 4.14.

Based on the specific consumption in charcoal-fueled pig-iron mills, it has been estimated that every year there are mined, produced, and transported more than 440,000 t quartz, 680,000 t limestone, 270,000 t manganese, and 27.4 million m^3 charcoal.

The ore is transported distances that vary on average from 10 to 100 km from mines to storage bins. Commonly, this transportation is by diesel trucks, but a small proportion is by train. The industrial limestone establishments that furnish this input to nonintegrated steel mills are often located at a radius of 200 km, from which it is brought by truck. The mines generally belong to independent owners who furnish the product to a series of industries. Silica and manganese are generally obtained not far from the mills and are transported by truck.

In the areas where mineral extraction and processing occurs, a series of environmental impacts are generated. The operations of transporting, handling, screening, and mixing with carrier agents causes the deposition and emission of dust and particulates in mining areas during transportation and near the steel mills.

During production, handling, and consumption of charcoal, a large volume of polluting fines are generated. The generation of charcoal fines amounts to about 25 percent of total production from the inception of its manufacture to its entrance into the blast furnace. This fines generation may be allocated as follows: 14.8 percent in charcoal kilns, 23.2 percent in loading and transportation, 25.2 percent in storage, and 37.6 percent in screening (Gomes et al. 1980). The large share of fines generated in the independent pig-iron production sector represents a significant wastage to the steel production process. Only a part of those fines generated in integrated mills are reutilized in the production process or sold to other industries, such as cement (CEMIG 1988:116).

Valuation of Water and Atmospheric Pollution Costs of Charcoal Fines

Although we have taken into account all sources of pollutants for which pig-iron manufacture is responsible, the present exercise will consider only the effects that result directly from charcoal production, transport, and handling from its inception to entrance into the blast furnace. Valuation here focuses on the environmental costs arising from the enormous quantity of charcoal fines produced and emitted by this activity.

As previously mentioned, around 25 percent of charcoal production is reduced to fines, the equivalent of around 1.8 million t charcoal dust per year.

> Some enterprises succeed in commercializing the charcoal fines produced in their mills to other mills (principally cement factories) that use them for combustion in their production process. However, the

> majority of fines generated in the independent pig-iron production sector is not reutilized in industry and becomes an unwieldy residual with severe polluting effects in the steel production process.
>
> *CEMIG (1988:233)*

It may be conservatively considered that around 65 percent, or 1.2 million t, of charcoal fines are produced annually as a residual and will directly pollute the soil, stream water, and urban areas near the mills. The final destination of this pollutant load is principally stream water flows, bringing as a consequence water pollution and increased turbidity, increasing the price of water treatment for human as well as industrial consumption. Such pollutant loads would be sufficient to pollute around 120 million m^3 of water up to a level of 1 percent of suspended solids. The environmental cost in this case may be determined by estimating the cost of water treatment.

Considering the cost of \$0.15/m^3 for treating this water, the environmental cost of production and deposition of charcoal fines from charcoal produced from around 650,000 ha of native forests and around 250,000 ha of eucalyptus may reach \$18 million. This implies that the environmental impact of each of the 7.2 million t of charcoal produced and consumed in steel manufacturing activities corresponds to a cost of \$2.50, as summarized in table 4.2.

INCORPORATION OF ENVIRONMENTAL COSTS IN CHARCOAL-FUELED STEEL MANUFACTURE

This discussion of environmental costs falls far short of expressing all damages caused by the production and use of charcoal in steel manufacture, as demonstrated by the fact that only six of the many identified environmental impacts are focused on. However, the impacts analyzed here refer to well-identified and physically quantified environmental damages characterized by the pertinent scientific and technological literature, and it has been possible to express their ecological effects in consistent economic terms, providing a rough order of magnitude to the associated costs.

This apology implies, on the one hand, that serious ecological damages such as species extinction, for example, were not taken into account here. This was so, first of all, because there were no quantitative data immediately available related to the dynamics of the populations of the species affected, and also because of the difficulties implicit in establishing relative importance among species and in calculating the economic value of the surviving members of a particular species. For that matter, it can therefore be asserted that *at least* the values identified for environmental costs represent the importance of the externalities and "free assets" that, once expressed in quantified monetary terms, might be considered as a basis to guide sustainable development in charcoal-fueled steel manufacturing.

From this perspective, the estimated values of the environmental costs associated with charcoal production and use in steel manufacture, as synthesized in table 4.2, are $57.00/t and $66.00/t of charcoal originating, respectively, from native forests and eucalyptus plantations. Considering the consumption of 875 kg charcoal for 1 t pig-iron, it was verified that the internalization of environmental costs would imply an increased value of $49.81/t pig-iron for charcoal obtained from native forests and $57.85/t for charcoal derived from eucalyptus plantations.

In 1995, from a total of 7.5 million t pig-iron produced in Brazil in charcoal-fueled pig-iron plants, around 2.8 million t were manufactured by charcoal from native forests and 4.7 million t by charcoal from plantations. The incorporation of the environmental costs in this national steel sector would involve an additional cost of $140 million for the first case, and of $272 million for the second, with a total of $412 million for the sector.

Once pig-iron is produced, it is impossible to distinguish the charcoal's origin. It thus becomes necessary to determine an average value for environmental cost associated with production of 1 t pig-iron, weighting charcoal participation according to its origin from native (48 percent) or plantation (52 percent) forests (ABRACAVE 1996). Accordingly, the average value of $54.00 should be considered appropriate for the environmental cost of national production of 1 t of pig-iron.

In recent years, 35 percent of national production of pig-iron produced in charcoal-fueled steel plants has been exported, or about 2.5 million t annually, following a growth trend. As a result of increased exports, the domestic price of pig-iron rose after reaching its lowest level in 1992 of $74.00/t. Since March 1994, pig-iron in international markets fetched a price of around $140.00/t. Despite this increase, it appears that the market has stabilized at this level; there has been no indication that pig-iron prices will return to their peak values in the short term (in 1989, international pig-iron prices attained $170.00/t). The principal factors that have influenced the trends in pig-iron prices have been scarcities of scrap iron, whose price rose to $144.00/t, and the withdrawal of the former Soviet Union from the international pig-iron market.

TABLE 4.15 COST OF PRODUCTION OF CHARCOAL-BASED PIG-IRON

Item	*Consumption (per t pig-iron)*	*Cost ($/t pig-iron)*	*Share (%)*
Charcoal	3.50 m^3	66.50	73.1
Iron ore	1.60 t	9.80	10.8
Manganese ore	0.04 t	0.56	0.6
Lime	0.10 t	0.45	0.5
Quartz	0.07 t	0.24	0.3
Electricity	60.00 KWh	1.85	2.0
Water	15.00 m^3	0.84	0.9
Refraction chemicals	1.70 kg	0.38	0.4
Replacement parts	—	0.17	0.2
Direct labor	8.00 hr	7.62	8.4
Indirect labor	3.00 hr	2.52	2.8
Total	—	90.93	100.0

From CEMIG, Instituto Brasileiro de Siderurgía (IBS), and the author.
For sales in the domestic market, $15.00/t is added by taxes; for exports, $10.00/t is added for freight charges from the factory to the port of Tubarão, Espírito Santo.

Just as scrap iron prices influence international pig-iron pricing, charcoal prices influence pig-iron on the domestic market. Among all cost components of pig-iron production, charcoal has the greatest influence, contributing around 70 percent of total cost. Table 4.15 shows the cost structure of charcoal-fueled pig-iron production. According to this table, the profitability of this industrial process depends fundamentally on the acquisition price of charcoal and, consequently, the incorporation of environmental costs involved in the production and use of this fuel can become a necessary strategy in discussions regarding sustainability of this economic activity. In the case in question, the production cost, incorporating environmental effects, of 1 ton of pig-iron for export would be nearly $140.00 FOB. At this level, pig-iron production would not be profitable in the short term, and the level of intensity of resource use would decrease.

CONCLUSIONS

From the identification and physical quantification of a series of environmental impacts caused by charcoal production and use in steel industry, it has been possible to identify monetary values, so as to permit discussion regarding incorporation of these costs in steel manufacture.

The apparent paradox of the lesser environmental cost attributable to charcoal produced from native forests arises from the methodological option to consider the primary environmental cost of land clearing and preparation (principally erosion) that follows deforestation as attributable to agricultural and ranching systems that are established in these areas, and not to deforestation for charcoal alone. This approach is based on an interpretation that frontier expansion for ranching and crop land in *cerrado* regions is inevitable. Thus, the use of all wood produced by deforestation for charcoal manufacture would result in a smaller environmental cost than that arising from plantation of energy forests.

Even with the methodological limits that hamper the valuation of important environmental costs such as, for example, the risk of species extinction, the results for those impacts that were evaluated imply that these costs are expressive. The estimated cost of environmental impacts at a national level resulting from charcoal production and use in steel industries, $412 million/yr, represents about 40 percent of the total value of pig-iron output. The total marginal costs of pig-iron for export, incorporating environmental values, would be about $140.00/t, a value close to the recent international market price. Were these prices to be internalized, the consequent loss in profitability would dissuade growth in charcoal production.

Discussions regarding sustainability of this economic activity should take into account the necessity to create mechanisms for compensatory reinvestment in the environmental area. Collection of forest taxes should respond to the environmental factors analyzed here.

5

Economic Valuation of Mangrove Ecosystems

MONICA GRASSO and
YARA SCHAEFFER-NOVELLI

The issue of ecosystem values and prices has been discussed for more than 20 years, since Odum (1971) published his work on the relationship between energy exchange supported by the ecosystems and the price of energy in the United States. These pioneering ideas made us think about all the services that are freely supported by nature and from which man can obtain goods.

The idea of placing a monetary value on nature sounds unethical and too materialistic, but every time that we make a decision we are in fact aggregating values. One can argue that value does not necessarily mean price. This is perfectly true, but only a few decision makers would be willing to justify their decisions based solely on judgment values: They need monetary figures to be able to negotiate in a world where money plays the major role. It seems that developing countries in particular suffer more from these effects because people are poorly educated and have little or no participation in decisions regarding the allocation of the natural resources on which they rely for their livelihoods.

A central problem in economics has always been to find ways to measure the value of economic activities and their related resources (inputs and outputs). Implicit to the idea of economic value, there are other values to be considered that can have a direct or indirect influence on the market price—for example, ethical/moral, cultural, or aesthetic values. Ethical and moral values are those based on the principle that all life forms deserve to live. Ehrlich and Ehrlich (1992) justify the existence of other species because they are the present expression of a continuing historical process of antiquity and majesty. Cultural values are intrinsic to different societies and customs and are not related to direct uses. Aesthetic values are related to ideas of beauty and the sensations that it engenders in human beings. Some of those values can be considered existence values and are usually implicitly aggregated to the market price.

We can say that the value of biodiversity to society will depend on the aggregated effects of the values described previously. Social preferences, technology, and the distribution of income and assets will also influence perceptions of value (Perrings 1995). According to the same author, valuation can be seen as a method of determining the relative importance of the environmental consequences of economic activities, helping environmental authorities make informed decisions about biodiversity conservation, where there are alternative ways of treating resources. All valuation procedures incorporate biases that result from socioeconomic factors, government policies and intervention, and lack of a developed market for some specific environmental service or good. The optimal use of natural resources cannot be achieved until there is a clear understanding of the reasons why misvaluation occurs, and of the rationales behind the need for accurate valuation methods (Bettencourt 1992). Misvaluation may arise as a result of (1) lack of defined property rights, (2) absence of established links between familial lineages and resource tenure rights, (3) open access to an ecosystem, and (4) informal markets.

Ehrlich and Ehrlich (1992) argue that the basic problem of conserving biodiversity is not likely to be solved until and unless a much larger proportion of the human population comes to share the view of the importance of Nature's existence. Species are major components of the life-supporting environment on which humans and their future welfare depend (Folke et al. 1992).

In this chapter we will give an overview of the status of the valuation of ecosystems, its importance and application, and a description of some currently used methodologies. An experience of mangrove valuation in Brazil will also be presented, with further discussion about the advantages and problems of using some valuation procedures in developing countries, compared to other experiences in developed countries.

ECONOMIC VALUATION AND GREEN ACCOUNTING

The gross national product (GNP), proposed by Simon Kuznets in 1941 as a measure of the size of the world economy, was not intended to become a measure of a country's welfare. Kuznets defined national income as "the net value of all economic goods produced by the nation," and his goal was to compare and differentiate economic and productive activities. The concepts of sustainability of environmental capital and the impact of economic activity on the environment are not taken into account in GNP. As Dieren and Hummelink (1979) state, our most serious mistake is in our method of calculating natural capital: Items deducted from nature are considered assets and are added to the GNP. This is the main reason so many people believe that they are benefiting from the exploitation of nature. The statement by the authors reflects how badly one can be mistaken in the naïve use of these concepts.

Today, many other accounting methodologies have been proposed—for example, SEEA (system of environmental economic accounts; Bartelmus and van Tongeren 1994); NNP (net national product; Daly and Cobb 1989); MEW (measure of economic welfare; Nordhaus and Tobin 1992); SNI (sustainable national income; Hueting 1980); HDI (human development index; United Nations 1995); HNA (human needs assessment; Max-Neef 1992). However, we still face the problem of how to value natural capital. There is little physical information about the true relationship between economic activities and environmental services. Alexander et al. (1996) argue that, conceptually, the contribution of an ecosystem to production and consumption could be assessed by asking how much production or consumption would be reduced given a small negative change in the amount of ecological services that are available to the economy.

For an environmentally adjusted accounting system to be reliable, it would be necessary to capture all the environmental services, account for the depreciation of both manmade and natural capital, exclude relevant categories of defensive environmental expenditures, and estimate damages that arise from economic activities (Lutz 1990).

Ecological Basis for Economic Valuation

Most valuation procedures are based on benefits generated by the ecosystem to mankind. It is important to bear in mind that valuation is restricted by factors such as knowledge about the ecosystem and biases present in economic models. Valuation seldom integrates all the linkages that the ecosystem services have with adjacent areas. Estimation of the total value of an ecosystem always involves complex problems related with system boundaries, scale, thresholds, and aggregation of values (Perrings 1995). The accuracy of the valuation will depend on the depth of knowledge about the ecosystem and its

TABLE 5.1 FUNCTIONS OF NATURAL ENVIRONMENTS (CONTINUES)

Regulation Functions

1. Protection against harmful cosmic influences
2. Regulation of the local and global energy balance
3. Regulation of the chemical composition of the atmosphere
4. Regulation of the chemical composition of the oceans
5. Regulation of the local and global climate (including the hydrological cycle)
6. Regulation of runoff and flood prevention (watershed protection)
7. Water catchment and groundwater recharge
8. Prevention of soil erosion and sediment control
9. Formation of topsoil and maintenance of soil fertility
10. Fixation of solar energy and biomass production
11. Storage and recycling of organic matter
12. Storage and recycling of nutrients
13. Storage and recycling of human waste
14. Regulation of biological control mechanisms
15. Maintenance of migration and nursery habitats
16. Maintenance of biological (and genetic) diversity

processes; Thus the first procedure for any valuation should be the identification of the ecosystem's functions and the goods it generates. Environmental functions are defined as the capacity of natural processes and components to provide goods and services that satisfy human needs—directly or indirectly (de Groot 1992).

For the systematization of the evaluation process, all functions can be classified into the following four function categories (mainly based on Braat et al. 1979 and de Groot 1992). For a more detailed description, see table 5.1.

1. *Regulation functions:* This group of functions is related to the capacity of natural and semi-natural ecosystems to regulate essential ecological processes and life support systems that, in turn, contribute to the maintenance of a healthy environment by providing clean air, water, and soil.
2. *Carrier functions:* Ecosystems provide space and a suitable substrate or medium for many human activities such as habitation, cultivation, and recreation.
3. *Production functions:* Nature provides many resources, ranging

TABLE 5.1 CONTINUED

Carrier Functions (Providing Space and a Suitable Substrate for:)
1. Human habitation and (indigenous) settlements
2. Cultivation (crop growing, animal husbandry, aquaculture)
3. Energy conversion
4. Recreation and tourism
5. Nature protection
Production Functions
1. Oxygen
2. Water (for drinking, irrigation, industry, etc.)
3. Food and nutritious drinks
4. Genetic resources
5. Medicinal resources
6. Raw materials for clothing and household fabrics
7. Raw materials for building, construction, and industrial use
8. Biochemical (other than fuel and medicines)
9. Fuel and energy
10. Fodder and fertilizer
11. Ornamental resources
Information Functions
1. Aesthetic information
2. Spiritual and religious information
3. Historic information (heritage values)

From De Groot 1992.

from food and raw material for industrial use, to energy resources and genetic materials.

4. *Information functions:* Ecosystems contribute to the maintenance of mental health by providing opportunities for reflection, spiritual enrichment, cognitive development, and aesthetic experience.

ECONOMIC VALUATION OF A MANGROVE ECOSYSTEM IN CANANÉIA, SÃO PAULO

Mangrove ecosystems are under all sorts of pressure, especially those of human origin: industrialization, urbanization, and agricultural development are some of the anthropogenic processes involved in mangrove degradation (Delgado 1981). Mangroves are found in intertidal areas throughout tropical and subtropical forests of the world, and until recently they covered almost 75 percent of the tropical coasts (Macnae 1974), growing in areas protected from waves and currents, in low energy systems. Like tropical forests of wholly terrestrial environments, mangroves have played an important role in the economies of tropical countries for thousands of years and constitute a reservoir and refuge for many unusual plants and animals (Hamilton and Snedaker 1984). The ecosystem supports both commercial and recreational fisheries by its capacity to export nutrients and provide many other direct and indirect services such as, for example, shoreline protection. Through their intricate root system, mangroves hold the muddy substrate, protecting the coast against the erosion produced by tides, waves, and even tropical storms (Delgado 1981).

Despite legal protection, mangroves have been subjected to many types of exploitation that restrict their capacity to support marine-estuarine life. In Brazil they are often the targets of real estate enterprises such as harbors and industrial zones, besides the widespread damage caused by oil spills and solid waste disposal.

Mangroves have important environmental functions that support many human activities. As Lal (1990) stated,

> They derive their values not only from fish and shellfish production and the economics involved in seashore protection, but also from certain subjective values, which may not have any direct market values. Recreational and aesthetic values are derived from the unusual mangrove flora and fauna populations, the protection they provide against soil erosion, mitigating floods, filtering nutrients and protecting interior lands from saline intrusion.

Some species that mangroves support are of direct and indirect economic and social value to human societies throughout the world (Saenger et al. 1983). Often, these extractable harvests are important not only for commercial uses but also for subsistence and thus may not pass through the formal market.

Examples of wetlands valuation include a study by Gosselink et al. (1974), who were the first to attempt to define monetary values for the products directly and indirectly provided by these areas, obtaining a value of about U.S.$82,000/ha/yr. Ramdial (1980) valued the socioeconomic benefits of the Caroni mangroves (Trinidad and Tobago) at about $8,000/ha/yr. Another study found approximately $9,900/ha/yr for Florida's mangroves based on their benefits for fisheries and tourism (Batie and Shabman 1978). In Brazil, the economic valuation of ecosystems arose with the necessity of providing compensation for different types of environmental impact. The first known mangrove valuation case was in Galinhos county (Rio Grande do Norte, Brazil), where Maciel (1986) established a value for the damages caused by the building of salt ponds. The total amount based on mangrove services lost with the project was about $53,757/ha/yr, a significant amount compared with other mangrove valuation studies.

To show the benefits of mangrove conservation, various approaches were employed assigning monetary values to resources that lack market price, showing both the advantages and disadvantages of each methodology. The purpose of this study is not only to assess the benefits of one such ecosystem but also to contribute to political decisions.

Bearing in mind that human necessities and aspirations vary according to culture, race, age, sex, seasons, climate, education, and income levels, it is important to characterize each cultural and socioeconomic setting when assessing and evaluating the importance of environmental functions to human society.

Characterization of the Mangrove Ecosystem

Mangroves occur along coastlines lying between latitudes 25°N and 38°S. Within this broad latitudinal band, they grow under a variety of climatic regimes. They also grow on a wide range of soil types, including heavy consolidated clays, unconsolidated silts, calcareous and mineral sands, coral rubble, and organic peats, with salinities close to 35 percent. The range of conditions in which mangroves occur naturally encompasses most environmental extremes except those where low temperature prevails.

Mangroves provide detrital nutrient input into coastal waters that support coastal fauna (Odum and Heald 1972). Moreover, mangroves are an important source of energy and nutrient flow into coastal water (Boto and Bunt 1981; Bunt et al. 1982). The variation in detrital productivity depends on mangrove plant species and environmental factors. The final fate of detritus depends on the particular food chain it has followed, which in turn depends on whether the detritus gets exported into the coastal waters or is trapped within the mangrove substrate. The extent of detrital export depends on the nature of the mangrove soil, the types of fauna present in the mangrove soils, the degree of tidal ebb and flow fluctuation, and the volume of water flow (Montague et al. 1987; Camacho and Bagarinao 1987).

The tides and runoff control the exchange of materials across the boundaries of the mangrove-estuarine ecosystem and are the major processes associated with the exchange of material with this system. How much water is transported between mangroves, estuary, and continental shelf is dependent on this potential hydrologic energy and the geomorphology of the region (Odum et al. 1979). These two factors also determine the extent to which the intertidal zone is inhabited by mangrove vegetation.

As mentioned, the development of mangrove swamps is the result of topography, substrate, and freshwater hydrology and tidal action. All these factors determine the resilience and the carrying capacity of this ecosystem. It is important to note that mangroves are already under stressed environmental conditions: The linkages between forest and external physical factors are very sensitive to any change. This type of ecosystem will support only a low degree of environmental impact. Once the vegetation is removed, it may not recuperate. For example, if the trees in a riparian mangrove are cut, the chances that a new tree will grow depend strongly on the intensity of the currents—if they are too strong, erosion will prevent recovery of the mangrove. It is also important to bear in mind that estuarine areas are very dynamic, subject to great change over a short period.

The hydrologic energy of riparian mangroves is high since they are dominated by river flow and tidal inundation, while fringe mangroves are influenced mainly by frequent tidal inundations. Basin mangroves have less hydrologic energy since they are found inland of fringe or riparian communities, and as a result they are less frequently inundated by either tides or river floods.

Leaf litter on the forest floor represents a major source of organic matter and nutrients that issue from mangroves to adjacent estuarine waters. Thus the balance of litter productivity, decomposition, and export influence the exchange of these materials at the boundary of the mangrove subsystem. It has been suggested by Pool et al. (1977) that litter production rates in mangroves vary with water turnover within the forest and the rank of the means of litter production (riparian > fringe > basin > scrub) (Twiley 1988). As the hydrologic energy increases, both the magnitude of the litter produced within mangroves and the proportion of this litter exported also increase.

The exact functional relationship between mangrove and fish fauna is a complex one. Nevertheless, the major primary producers in all these linkages are mangrove plants, which form the basic energy source for the interface between land and water ecosystems.

Mangroves and wetlands overall have always been statically associated with fishery yields. The life cycle of the penaeid shrimp begins in the open sea as eggs. After drifting in the pelagic larval phases, the postlarvae enter estuarine areas on flood tides, and they seek substrates, such as mangrove roots, to which they cling until the next tides, when they penetrate deeper into the estuary (Bell 1989). According to Turner and Boesh (1989), shrimp live a benthic (bottom) existence while they grow in the estuary, an environment that offers food and refuge from predators. Shrimp are first-order consumers of detritus (Day and Yanez-Arancibia 1987) and detrital material is the result of the net production of the wetland areas.

Mangrove habitat provides increased physical structural complexity that will decrease the efficiency of predatory fish in feeding on juvenile shrimps and fishes. The close relationship between penaeids and mangrove habitats in the tropics is evidenced by significant correlation between the estimated maximum sustainable yield (MSY) of penaeids and the area of mangrove habitats (AM) in several regions of the world (Macnae 1974; Turner 1977; Martosubroto and Naamin 1977; Staples at al. 1985). In a recent reworking of most of these data, Pauly and Ingles (1986) showed how the following relationship,

$$\log_{10} \mathrm{MSY} = 2.41 + 0.4875 \log_{10} \mathrm{AM} - 0.0212L,$$

where L is degrees of latitude, explained 53 percent of the variance in the dependent variable (MSY).

Robertson and Alongi (1992) state,

> While such regression models offer appealing evidence of a link between mangroves and commercial fisheries they have a variety of analytical and theoretical drawbacks. The MSY data is itself questionable and subject to a high degree of error like most fisheries data. Furthermore, even if the data are assumed to be corrected, nearly half the variance in MSY is not explained by area of mangrove or latitude, and is caused by other factors. In addition, three hypotheses might explain the cause of a correlation between mangrove area and commercial penaeid yields. First, mangrove waterways act as nursery grounds for penaeids, which as they mature, move offshore and enter the commercial fishery. Strong evidence in support of this hypothesis comes from the many surveys of postlarval and juvenile penaeids in near-shore habitats which show some penaeids to be mangrove-associated as juveniles. However, as pointed out earlier the causal link between these juvenile penaeids and mangroves has not been established experimentally. In addition, a large proportion of the penaeids contributing to commercial catches are not mangrove associated as juveniles.

Fish species composition and richness in any tropical mangrove system will depend primarily on (1) the size and diversity of its habitats, with their distinct flood and tidal regimes, (2) the proximity of mangroves to other systems, and (3) the nature of the offshore environment, particularly depth and current patterns (Bell 1989).

Proximity to other mangrove systems ensures colonization by even those species with no or short larval duration, and by movements of adults and juveniles. A corollary is that proximity to nonmangrove areas, such as coral reefs, may influence fish species composition in the mangrove.

Description of the Area

Cananéia is a community on the south coast of the state of São Paulo, 25°S (Brazil), with an approximate area of 10,000 hectares (ha) (Besnard 1950). The mangrove area, about 5,588.6 ha, is nearly equally divided between what are considered "high forest" and "low forest," and it accounts for over one-third (37.6 percent) of all mangroves present in the state of São Paulo (Herz 1990). Furthermore, Cananéia is part of a system called the Complexo Estuarino-Lagunar de Iguape e Cananéia (Estuarine and Lagunar Complex of Iguape and Cananéia), which has two main settlements, Cananéia and Iguape (figure 5.1), encompassing a total mangrove area of approximately 13,305 ha (SMA 1990).

The typical mangrove species on the area are *Rhizophora mangle, Laguncularia racemosa,* and *Avicennia schaueriana.* From an ecological point of view, mangroves are the main component of the coastal lagoon. They are part of an estuarine environment in which they play an essential role in the high productivity of the ecosystem, from both an ecological and economic point of view.

This region is considered reasonably preserved, being classified by the International Union for Conservation of Nature and Natural Resources (IUCN) as the third estuary of the world in terms of primary productivity (Adaime 1985). It has only 51 ha of altered mangroves and 215 ha of degraded mangroves (Herz 1990).

The population of Cananéia includes about 7,726 people, who depend primarily on subsistence fishing and agricultural activities, with tourism becoming responsible for an increasing share of the local economy. Local residents have used the mangroves as sources of energy and building materials and for fishing purposes. More than 80 percent of the fisheries activity occurs in the inland waters because

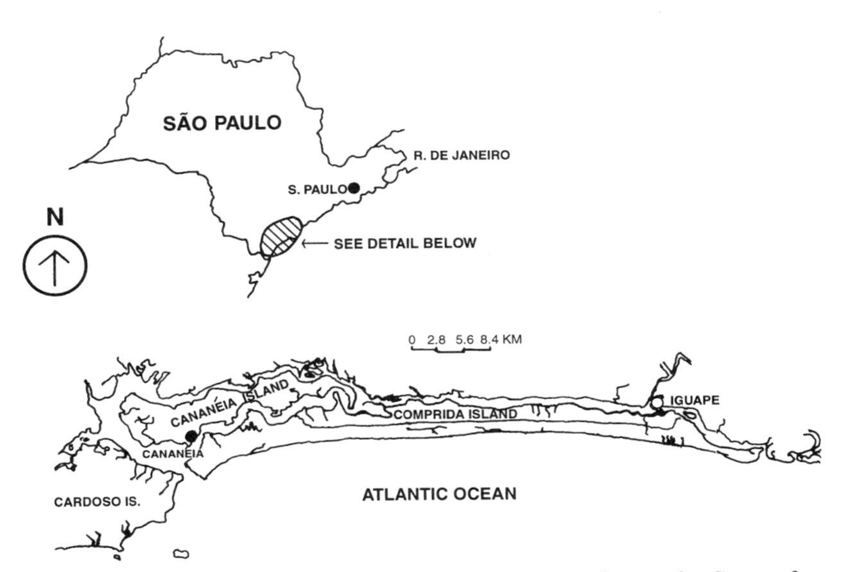

FIGURE 5.1 Map of the Iguape-Cananéia Lagoon Complex on the Coast of São Paulo, Brazil

of the abundance of such river and lagoon species as mugilidae and clupeidae. The most commonly exploited products are fish and shellfish.

The cultural value of such forests for those living in the area is difficult to assess objectively, but mangroves are part of the landscape and of their lives. This ecosystem also interests tourists. Developers have recognized the recreational potential of the area, one in which beaches are combined with extensive mangrove forests within a shallow and peaceful coastal lagoon.

Economic Valuation of the Ecosystem

The economic analysis of mangrove values requires the identification of their function and resources; an assessment of their importance in terms of the impact of mangroves on, or relevance of mangroves to, economic activities; and the value of the nonmarket functions they serve. We have followed five steps described by Barbier (1991):

1. Define the wetland area and specify the system boundary between this area and surrounding regions.

2. Identify the functions, attributes, and structural components of the wetland ecosystem and rank them in terms of importance (e.g., high, medium, low).
3. Relate the functions, attributes, and structural components to the type of use value (e.g., direct use, indirect use, and nonuse or preservation).
4. Identify the information required to assess each function, attribute, and component, and how to obtain this data.
5. Use this information to quantify economic values, where possible.

The valuation of the mangrove was based on its regulation, carrier, production, and information functions, and for each of these we chose the best methodology to be applied based on the socioeconomic characteristics of the area of study. For each mangrove good and service, an economic valuation methodology was proposed. For simplicity of the valuation analysis, the mangrove's functions were divided into direct and indirect use and existence values. The direct use was evaluated by fisheries, sport, commercial, and artisanal activities; for the indirect use, we evaluated all other services and goods that are not traded in the formal market; finally, the existence value was based on peoples' expressed willingness to preserve the ecosystem for future generations.

Valuation of the Environmental Function (Nonmarket)

EXISTENCE VALUE: THE CONTINGENT VALUATION APPROACH Contingent valuation is a survey-based methodology (CVM) that may be used to discern the value of a natural area lacking data on market or surrogate-market prices, and it is frequently used to reveal existence values. It involves the direct questioning of consumers to learn how they would react to certain situations. The respondents are presented with material, often in the course of a personal interview conducted face-to-face, which consists of three parts (Mitchell and Carson 1989):

1. A detailed description of the good(s) being valued and the hypothetical circumstance under which it is made available to the respondent;
2. Questions that elicit the respondents' willingness to pay for the good(s) being valued;
3. Questions about respondents' characteristics (for example, age, income), their preferences about the good(s) being valued, and their use of the good(s).

Contingent valuation methodology relies on standard neoclassical economic principles and uses one of two hicksian measures of consumer's surplus: compensating variation and equivalent variation. Compensating variation, here called willingness-to-accept-compensation, corresponds with the payment or change in income necessary to make an individual have no preference between an initial situation and a new situation in which different prices prevail. Equivalent variation, or willingness-to-pay (WTP), may be viewed as a change in income equal to a gain in welfare resulting from a change in price.

These approaches remain controversial, since CVM is primarily based on hypothetical situations and depends on some "reference operative conditions" (Cummings et al. 1986): Respondents must be familiar with the service to be evaluated; they must have some experience with consuming this service, and there must be little uncertainty regarding the "contingent" situation described.

McConnell (1988) described some reasons why CVM is widely used, and how some restrictions can make it a reliable method. There is work that supports the evidence that the hypothetical and actual compensating variations are not greatly different; several studies show that the compensating variation estimated from behavioral models is similar in magnitude to that obtained from contingent valuation studies; and finally, there is evidence from research involving experimental exchanges in laboratories that people generally can (and will) tell approximately what they are willing to pay.

It is important to observe that all of these works are based on studies done in developed countries, primarily in the United States, where the methodology was created. In developing countries, the situation will be much more complicated and uncertain. First, not all the "reference operative conditions" will be satisfied. Usually, in locations where the ecosystem has an important role in community life, there will be almost complete lack of information about market prices and ecosystem services, coupled with very low incomes. Therefore, for the application of this methodology in developing country cases, it is necessary to correct and adapt the methodology to the local situation. For example, Grasso (1994) substitutes the monetary value for the goods that fishermen would have to purchase if they could no longer fish in the mangrove area—for example, a boat engine that would enable them to fish in open sea. Using this method was necessary because the fishermen were not able to assess values associated with their way of life, the only monetary income they perceived being the surplus from their catch, most being for subsistence.

Four formats of contingent valuation are (1) the direct open-ended method, (2) the bidding game, (3) the payment card, and (4) the take-it-or-leave-it method (for more details on these methods, see Mitchell and Carson 1989). It is important to have an iterative process offering respondents time—and a learning opportunity—to develop an attitude toward paying for goods. Consequently, it has been argued that bidding games will encourage respondents to arrive more accurately at their maximal WTP value (Cummings et al. 1986).

The application of the CVM in this study was based on the referendum model format. The referendum model was used because it is the valuation method furthest from an attitudinal study, since it forces a choice on the respondent, rather than probing the respondent's attitudes (McConnell 1988).

WILLINGNESS-TO-PAY This study applied the WTP methodology by direct survey with the local visitors. The formulation of the questionnaire was based on ten fixed values ($1; 2.50; 5, 10; 20; 35; 55; 80; 110; 150; 200) given by a simple referendum made previously. Respondents had to answer about their willingness to pay for the preservation of the mangrove ecosystem in Cananéia. The methodology works with qualitative response models (QR), most frequently using discrete dependent variables, where the respondents have only two options for an answer, yes or no (referendum model). When respondents answer yes, we can infer that they are better off with the preservation of the ecosystem and with less income. When the respondents answer no, they will be better off keeping the money for themselves. We assumed that socioeconomic and environmental characteristics explained part of the individual's answer, and they were gathered in a vector X, so that

$$\text{Prob}\,(\text{WTP} = 1 = \text{yes}) = F(\mathbf{b'c}), \text{ and}$$

$$\text{Prob}\,(\text{WTP} = 0 = \text{no}) = 1 - F(\mathbf{b'c}).$$

The data obtained were analyzed, using the logistic[1] distribution, for the following logit function:

$$P(\text{WTP} = 1) = 1/[1 + \exp(-a + b_1X_1 + b_2X_2 + \ldots + b_nX_n)],$$

where

$P(\text{WTP})$ = probability of paying,
a = constant,
$b_1 \ldots b_n$ = coefficients, and
$X_1 \ldots X_n$ = individual characteristics variables.

TABLE 5.2 RESULTS FROM THE CONTINGENT VALUATION STATISTICAL ANALYSIS

Variable	*Coefficient*	*Std. Error*
Constant	2.047	0.29
Income (Y)	0.2	0.007
Bid	−2.57	0.0006
Level of knowledge (LKM)	0.58	0.26
Capacity of supporting crabs (CSC)	0.77	0.27

Many variables were tested, but the ones that demonstrated the best correlation were income (Y), the level of the knowledge of the mangrove (LKM), and its capacity to support crabs that have commercial value (CSC), so we can say that the willingness to pay for mangrove preservation depends on

$$\text{WTP} = f(\text{Y}, \text{LKM}, \text{CSC}).$$

Table 5.2 presents the results obtained in the statistical analysis. Only approximately 16 percent of the people surveyed agreed to pay. The average monthly income for these people was $1,767. The maximum and the average amount that they were willing to pay (calculated based on the logit function) were, respectively, $134.90 and $43.85 per year. Making a simple aggregation to the area of mangrove in the region, we will have a value of approximately $230 to $710/ha/yr for the ecosystem.[2] This value represents only the perceived value to the visitors, not including the value of the functions of the ecosystem.

We were looking for a monetary value associated with the overall quality of the environment. This type of assessment can, however, lead to an error called part-whole bias, or desegregation effect, in which respondents value an entity different from the intended commodity. Such a bias would arise if respondents valued environmental quality, all endangered species, or the total value of a species when it is no longer endangered (Boyle et al. 1994). In each of these cases, respondents would be valuing a larger "package" than the one intended by the researcher, so contingent valuation estimates would not distinguish between parts of a valuation issue and the issue as a whole.

For Mitchell and Carson (1989), the part-whole bias can arise from either fundamental flaws in the design of a survey instrument or from the inability of respondents themselves to answer contingent valuation questions, because respondents generally do not have choice experience or knowledge of the object being evaluated. That is why it is important to test how strong the relation is between user and environment. Furthermore, to minimize the potential biases, we need to include a description of the commodity that we are evaluating; the inclusion of maps and photographs is also proposed.

Smith (1992) argues that the framing of the commodity to be valued using CVM must reflect an understanding of how people perceive it, what people consider to represent related goods (either complements or substitutes), and how people understand the processes involved in altering their consumption patterns as part of adjusting to an exogenously imposed change in the good of interest.

Sometimes respondents have difficulties providing a response to the initial contingent valuation question posed. However, once respondents answer an initial question, responses to subsequent questions that enhance or diminish quantity or quality may follow an expected pattern, because respondents recognize that the initial quantity has been changed. This aspect was used by Whittington et al. (1991b) when working in developing countries. The authors found that when subjects have limited education and monetary resources, when we give them more time to think, the bids are usually lower. That means that when one applies this technique in developing countries, one should consider some aspects that will probably end up increasing the financial expense of the study (such as the necessity of personal interviews). We found that people in Brazil usually had problems in answering the CV questions. They are not accustomed to associating values with natural resources for the following reasons:

1. It is uncommon to pay (directly) for the use of an ecosystem in Brazil, and when there is a fee, the amount is very low (unlike in the United States, where one pays to go to parks and to fish). That aspect can be seen in the marketplace, where individuals cannot always recognize the value of a product because it is very cheap.
2. They do not have information about the functions and the ecological importance of that resource.
3. Because Brazil is considered to be a country rich in natural resources, people seem to idealize inexhaustible sources of raw materials.
4. People also have difficulties understanding the connection between economic loss and depletion of natural resources.

These are some aspects that justify the necessity of a personal interview with the users: if we did not use this procedure, there would be too many incomplete answers. The increase of the frequency of "no" responses and the potential importance of strategic considerations linking the user's informed values to their stated value would limit usefulness of CVMs in developing countries (Whittington et al. 1991a).

Looking at the income of the people and the value that they are willing to pay for the improvement of the environment in both cases, we note that, despite their low incomes, Brazilians are indeed willing to pay ($134/yr). Bockstael et al. (1989) obtained $183/yr applying CVM to the Chesapeake Bay (United States). The income of the interviewees was almost 4 times greater than those in Brazil (table 5.3); thus we can conclude that the Brazilian's bids were very high for their standard of living. This bias may arise from the aspects highlighted previously and because different CV formats have, predictably, different performance characteristics (Hoehn and Randall 1987). The allocation decisions by households are the result of a complex decision process involving the household's monetary and nonmonetary resources.

It seems that the most striking implication from the study was the difficult task of valuing marginal changes in a natural resource, when those changes represent small proportions of the total environmental assets in question. Part of this difficulty may arise from the public's inability to appropriately assess small proportionate shifts in their welfare as shown by the risk literature and the respondents' general lack of familiarity with the commodity being evaluated.

Kaneham and Knestch (1992) found that there is an order effect in WTP responses when the values of several goods are elicited in succession. The same good elicits a higher WTP if it is first on the list rather than valued after others. That is why one must be careful with the proper formulation of questions. Probably the best way to apply CVM is first to test whether this kind of bias is likely to critically affect the results.

In any case, CVM is an important model to reveal consumers' behavior and preferences, and it is a powerful and versatile tool for measuring the economic benefits of the provision of nonmarketed goods. It is potentially capable of directly measuring a broad range of economic benefits for a wide range of goods, including those not yet supplied, consistent with economic theory. Through this methodology, we also infer socioeconomic aspects of a population: the level

Table 5.3 Comparison between Case Studies (Chesapeake Bay and Cananéia)

	Chesapeake Bay (US)	*Cananéia (Brazil)*
Evaluated aspects	Water quality (20% reduction of nitrogen and phosphorous)	Ecological functions of the mangrove ecosystem
Ecosystem area	498,000 acres of tidal and nontidal wetlands	55,886 m^2 mangroves (28,655 m^2 high forest; 27,231 m^2 low forest)
Visitors' main activity	Recreational (swimming)	Recreational (fishing)
Number of interviews	959 residents	483 tourists
Survey method	Telephone	Personal interview
Type of users	Residents of Baltimore, Washington SMSA	Tourists (mainly from São Paulo)
Average income of respondents	Whites: $40,000/yr Nonwhites: $25,000/yr	$12,000/yr —
WTP (users)	Whites: $183/yr	$134/yr
	Nonwhites: $34/yr	—
Variables that explain the WTP of the respondents	Race (income)	Knowledge about mangroves and their capacity to support crabs, which have commercial value

of education, income and their relation with and knowledge about the environment that is being evaluated.

Recreational Value

TRAVEL COSTS The travel costs approach (TCM) is another means to value unpriced goods. In developed countries, this approach has been extensively used to derive a demand curve for recreational facilities and open space resources (Clawson 1959). Outdoor recreation is a typical example of such an unpriced good. Although the approach has been associated solely with recreational analysis, there is no reason it could not be applied to a broader range of problems (Sherman and Dixon 1990).

The TC approach enables us to calculate a demand curve for unpriced goods and identify the socioeconomic characteristics of the visitors. In cases of environmental damage, such as beach contamination during the holiday season, TC establishes the cost of damages experienced by users. These are important tools for the establishment of management policy and environmental compensation. By measuring the level of tourism exploitation in natural areas, TC analysis can also determine whether a given level of tourist activity may be harmful to the values expressed by users. Under the assumptions used in the analysis, the benefits calculated must be considered as a minimal estimate of the total benefits derived from a recreational facility.

In this project, the TC technique was used to determine the consumers' hypothetical valuation of the recreation activity in Cananéia (SP, Brazil). About ninety-five families were interviewed to determine the following:

1. The travel distance, calculated on the basis of road distance (in km) from place of origin to Cananéia.
2. The total population of each center of origin. These data were taken from the 1980 nationwide census (IBGE 1980).
3. The population characteristics. Many socioeconomic characteristics of the families were identified, and those with the greatest degree of statistical correlation to the visitation data were selected as part of the estimation model.

The most important factors analyzed were based on the various costs involved in visitors' trips (costs of travel time, imputed costs of visitation time, costs of transport, and the total costs of the trip) and their relation with the sport activity, visitation rate, time of visitation, travel distance, level of education, and knowledge of the ecosystem.

We imputed opportunity costs of travel time based on one-third of a visitor's monthly salary (Cesario 1976; Pearce 1976), which demonstrated good statistical fit with the results. Some authors propose other means to measure the time value: Pearce (1976) and Smith (1983) use the full salary; other authors use different proportions of total monthly salaries. More detailed analysis of the time costs of the trip can be found in McConnell (1975), Bockstael et al. (1989), Mendelsohn (1987), and Smith (1989).

The economic analysis was based on different costs involved in the trip (trip costs, time costs, hotel and transportation costs) related to the following variables: type of activity, frequency of visits to the area, time spent at the site, distance traveled, level of knowledge about the ecosystem, and education.

To facilitate the classification of visitors according to trip expense, travel distances were divided into ranges:

Zone A: 0–100 km
Zone B: 100–300 km
Zone C: 300–500 km

Distances beyond 500 km seldom appeared in the survey sample.

The visitation rate for Cananéia was estimated based on the Sherman and Dixon (1990) equation:

$$\text{Visits}/1{,}000\ \text{people}/\text{yr} = [(V_i/n)N \times 1{,}000]/P,$$

where

V_i = visitors from zone i
n = number of interviews = 95
N = number of visitors/yr = 34,212
Visits = frequency of visitation/yr
P = total population in zone i

To obtain the consumers' surplus, it was necessary to establish a demand curve. The demand curve was plotted based on the following linear relationship of the coefficients, given by

$$\text{Visits} = 3.7313081\,(\text{costs}^{-0.6305822}).$$ [3]

We used only the recreational fisheries data because we intended to reveal the monetary relationship between mangroves and the rate of visitation. The average annual income of the tourists' families was $12,000 and approximately 66 percent of them had a graduate degree. Exactly half the tourists came to Cananéia to make a boat trip inside the mangroves, and about 26 percent of the visitors were there because of the recreational fishery and were willing to spend $120/day in their activity. Based on the demand curve, we obtained a consumer surplus of $2,720/visit/yr, giving an approximate aggregate value of $3,583/ha/yr for the mangrove ecosystem. We can assume that this value corresponds to the monetary value associated with the mangrove in supporting the recreational activities mentioned previously.

Indirect Valuation of Environmental Functions (Ecosystem Value)

The values described in this section are based on the capacity of the ecosystem to maintain a certain functional level of inputs and outputs. Many environmental functions are interlinked, and each function is the result of the interactions between the dynamic and evolving processes and component structures of the total ecological subsystem of which they are part. When considering the values associated with the use of environmental services, it is first necessary to define what level might represent a sustainable intensity of use. Although the methodologies that follow were not actually applied in the Cananéia case study, they should be taken into account if one expects to discern the values associated with ecosystemic functions.

SURROGATE MARKET PRICE The surrogate market price (SMP) is based on the principle that we can estimate the value of an environmental good or service by examining the price paid for a closely associated good that is traded in the market. This approach uses observable market prices for one good to estimate the value of an environmental good that does not have a price of its own. This methodology can be used to find the value of the following services.

Maintenance of Water Quality (Tertiary Treatment per Nutrient Sink and Cycling) The removal of nutrients improves water quality and helps to prevent eutrophication, precluding the need to build water treatment facilities. Under certain circumstances, mangroves can be used for treatment of domestic wastewater from small, nonindustrial communities. When mangroves remove nutrients (and pollutants), they are called sinks. This function is particularly valuable in disposal of nitrate pollutants, which can be converted to gaseous nitrogen and circulated back to the atmosphere as a result of denitrification. When materials are exported, mangroves act as sources (Twiley 1988).

The wastewater treatment function can be assessed based on data concerning the capacity of mangroves to remove nutrients from adjacent waters. A value equivalent to this level of nutrient retention can be derived for the mangrove area based on the regional costs of waste treatment.

Erosion Control, or Soil Retention Mangroves can stabilize shorelines by reducing the energy of waves, currents, or other erosive forces, while their roots simultaneously hold the bottom sediment in place. This can prevent the erosion of valuable agricultural and residential land and associated property damage. Occasionally, they may actually help to build up land. Sediment is the major water pollutant in many river systems. Mangroves commonly occupy basins and can act as pools where sediment can settle. Although the buildup of too much sediment in a wetland may alter its biological functions, floodwater storage, and ground water exchange capacities, the quality of ecosystems downstream will be protected if suspended sediment is retained in the headwaters. Retaining sediment in headwater wetlands will lengthen the life span of downstream reservoirs and channels, and reduce the need for costly removal of accumulated sediment (Dugan 1990).

Studying how mangroves contribute to erosion control requires the determination of the most suitable way to measure sediment capture, using techniques from watershed hydrology. It may also be possible to construct an "artificial mangrove" at important points along the riverbank to observe how sediment trapping occurs. The value of this function will be based on the costs of an equivalent function: dredging the same amount of sediment.

Disturbance Regulation (Storm Protection and Flood Control) Through their capacity to store precipitation and release runoff evenly, mangroves can diminish the destructive onslaught of flood crests downstream. Preserving natural storage can avoid the costly construction of dams and reservoirs. To learn the effect that an individual mangrove area may have on downstream flood flows would require a detailed hydrological study of each site. However, it appears that the effect of wetlands in the upper watershed ($1^{st,}$ $2^{nd,}$ and 3rd order streams) is discerned only in their effects on areas a few kilometers directly downstream. In contrast, large wetlands on the main stream (5th order) have a substantial effect on flood crests much farther downstream (Dugan 1990).

Hurricanes and other coastal storms cause wind damage and flooding. In the developed world, the principal consequence is property damage; in poor tropical nations it may be human injury and death. Mangroves help dissipate the force and lessen the damage of coastal storms. In recognition of this important function, measures are being taken to protect and restore coastal wetlands in many countries (see Farber and Costanza 1978).

The measurement of flood control will be based on data already available on the quantity of water that mangroves can retain in their sediment. We can then calculate the value of this function based on the current price of clean water in the region. The capacity of mangroves to protect adjacent areas from storms can be measured by the difference between the wind intensity in the open area of the forest and that in the protected area, and by estimating the value of the possible damages if the houses close to the area did not have the mangrove to protect them.

BIODIVERSITY—A QUALITATIVE VALUATION The economic valuation of biodiversity is still very difficult and generates a great deal of controversy. Ehrlich and Ehrlich (1992) found two practical problems with assigning values to biological diversity. The first is a problem for economists: Discerning the true economic value of any particular segment of biological diversity is not possible. We do not know enough about any gene, species, or ecosystem to be able to calculate its ecological and economic worth in a larger scheme. The second practical problem is one for conservationists: Many species do not seem to possess any value as conventionally measured. Unfortunately, the species whose members are fewest in number are those most likely to become extinct. Calculating the densities of individuals that occupy the ecosystem area and recognizing their species representation can give us an idea of the level of diversity that the ecosystem can support.

Direct Valuation of Ecosystem Goods and Services Based on Market Prices

The accounting of ecosystem productivity is an adaptation of the change-in-productivity approach (Sherman and Dixon 1990), based only on the net annual value of fish harvests that are dependent directly or indirectly on the mangrove ecosystem. The main purpose of using that methodology was to obtain the relation between fish capture and mangrove area.

In our study, this methodology was based on the annual catch of commercial boats fishing in the region (i.e., valuation of commercial fisheries). To calculate the productivity value of the ecosystem, the estimated yearly production under sustainable fishery harvest is estimated at the current market prices. For those products that neither are exploited nor have a market price, using the surrogate market price or another approach may be possible, depending on the characteristics of the resource. The first step in the application of this methodology is to prepare a comprehensive list of the fish and shellfish that depend directly and indirectly on the ecosystem to be evaluated, based on data and references available.

The greatest difficulties in valuing fishery potential arise from the fact that aquatic resources belong to the category of common property goods (such as wildlife and natural landscapes) and that fisheries cannot be measured within a limited region: Fish are often landed in areas far from where they were caught. In our case study, to account for productivity by mangrove area, we selected only the yield of the species whose life cycles depended on this ecosystem. The calculation was based on the commercial data of the fish harvested, obtained from the São Paulo State Supply Company (CEAGESP, a food wholesaling enterprise) from 1974 to 1989. We also interviewed the crews of fishing boats to determine the approximate capture radius and the net returns from their activity. The costs accounted for in measurement of net returns were fleet investments, fuel, and general crew expenses. For each boat, we obtained data on the number of workers, type of gear, fish caught, area of harvest, and monthly yield. All boats using the ports of Cananéia were catalogued. The relationship between mangrove area and yield can be calculated if there are enough historical data for both variables. Usually we expect to have a growth in catch as fishing effort increases. This will occur up to a maximum point (maximal sustainable yield), after which a decrease is anticipated in the catch even with an increase in effort, because of the depletion of the stock (Lynne et al. 1981). The total value of the area of the mangrove was calculated based on the market price of the catch and on the fishing area, as suggested by Moulton (1991).

Based on the average quantity of fish harvested (170 t /yr) and the commercial price of each species caught, we obtained an approximate gross value of $7,357,041/yr. The total amount spent on the activity was about 15 percent of the value of total production, resulting in a net value of $6,253,484/yr. Since the approximate area of mangroves within the area of capture is 13,305 ha, we obtained a value of $470/ha/yr for the mangrove.[4]

The evaluation of mangrove benefits to the fisheries economy failed to recognize the principle of with-and-without analysis (Hufschmidt et al. 1983). Specifically, since the fish associated with mangroves are not unique to that ecosystem, the assumption that without the mangrove there would be no fish is not valid, as some will be provided by other ecosystems such as those based on seabeds and coral reefs (Lal 1990). Thus there is some bias that we could not control, such as the total amount captured that could be attributed to the mangrove area in question.

Socioeconomic Function

The household production approach (HPA) has been used by economists to study many issues concerning farm households. These include labor supply, transportation, intrahousehold decision making, and so on. The same framework can be used to study the benefits provided by an ecosystem to a community.

In our case study, we used this approach to measure the portion of fishermen's welfare that resulted from the existence of the mangrove ecosystem. Artisanal fishermen keep a close relationship with the mangrove, so gradual damages borne by the ecosystem are clearly reflected in their lives. Traditional methods of fishing for subsistence are still being used, some with minor changes. Fishermen continue to live in small villages close to the rivers, harvesting for their own consumption and selling what remains of the production. There is not a clear distinction between subsistence fishing and small commercial artisanal fishing; both use similar methods of harvesting.

At the time of our study, Cananéia had 768 fishermen, most of whom (54.3 percent) live in the urban area after migrating from rural zones (SMA 1990) because of real estate speculation and the decline in agricultural activities that results from lack of economic incentives. Fishermen who live in the rural zones have two distinct types of economic production:

1. Fishing and agriculture. Fishing is a seasonal activity, restricted to the harvest of some species. Harvesting is accomplished by groups, usually with low-cost equipment.
2. Artisanal fishery. Fishing goes on all year. The fishermen usually have a good knowledge of the species' life cycles and they use more advanced equipment. Ownership of the equipment is fundamental for productive organization and defines the difference between boss and employee in dividing up the production (SMA 1990).

We can define a simple household model that describes the utility function of the fishermen in Cananéia as follows:

$$U = U(G_m, G_s, F, L),$$

where

G_m = vector of market goods purchased by the household
G_s = vector of subsistence goods purchased by the household
F = vector of minor forest products purchased by the household
L = vector of leisure activities

Two of these goods, G_s and F, are both produced and consumed by the household. The household uses the surplus from fishery production and the other products to buy market goods.

The harvest is derived from the lagoon (51 percent), rivers (30 percent), and coastal waters (19 percent). The fact that more than 80 percent of the harvest is obtained from inland water bodies can be explained by the lack of a seafaring tradition among the population (Mourão 1971), by the abundance of the river and lagoon stocks, and also by the difficulty of fishing in open sea (requiring more advanced equipment and boats). Only 12 percent of the boats have an engine and 45 percent of the fishermen do not have their own boat; 75 percent have their own nets, 15 percent use wires, and 21 percent use other equipment (SMA 1990). The majority of the production is bought by CEAGESP. Besides fishing, there is also harvesting of oysters and crabs, which is illegal because of overexploitation, having nearly led to the extinction of these species.

The most important species to the artisanal fishermen are bluefish (*Pomatomus satator*); catfish (*Bagre* species, *Genidens genidens, Sciadeichthys luniscutis, Notarius grandicassis, Arius spixii, Netuma barba*); southern king croaker (*Menticirrhus americanus*); Atlantic croaker (*Micropogonias furnieri*); large head hairtail (*Trichiurus brasiliensis*); strot (*Etropus crossotus, Syacium papillosum, Achirus* species); mullets nei (*Mugil platanus* and *M. curema*); weakfish nei (*Cynoscion leiarchus* and *C. acoupa*) and snoaks nei or robalo (*Centropomus* species); white shrimp (*Penaeus schimitti*); crabs (*Callinectes* species), and oyster (*Crassostrea* species).

For the valuation approach, approximately 220 fishermen were interviewed about their socioeconomic situation and net returns from the fishery activity (tables 5.4 and 5.5). We obtained information about their families and data about their fishing performance. Net revenue was calculated from their gross monthly revenue minus their fishing expenses. The costs of the fishery activity include all taxes paid, boat investments, maintenance and amortization of the equipment, and costs involved with routine daily activity, such as, for example, fuel and food for the crew. The net monthly revenue that the fishermen obtain from the mangrove ecosystem was estimated to lie between \$119 and \$157. The aggregate value attributed to the mangrove was low (approximately \$228/ha/yr), because of the poverty of fishermen in this area. Expenses of fishing equipment never exceeded \$16/mo. The average number of people in the fishermen's family is six and most of them work in fishing activity, with an average expense of \$124/mo.

TABLE 5.4 SOCIOECONOMIC SITUATION OF THE FISHERMEN IN CANANÉIA

County	*Fishermen Interviewed (N)*	*Average Monthly Income (from Fishing Only)*	*Total Monthly Income*	*Average Monthly Fishing Expenses*	*Average Monthly Household Expenses*
Centro	108	$157	$205	$8	$166
Carijo	127	$128	$143	$16	$125
Acarau	20	$127	$137	$6	$120
Morro	48	$155	$196	$10	$110
P. Cubatão	88	$119	$149	$9	$102

The low quality of life of fishermen in the coast of São Paulo seems to have no simple explanation, since the fish are sold in São Paulo and other cities for a high price. Nevertheless, failures in fish capture, distribution of the harvest, and the high cost of production in an unstructured enterprise are the causes of the primitive conditions and the misery faced by fishermen, and of the relatively small role that the fishery plays in the Brazilian economy.

This work revealed the importance of the mangrove to the artisanal fishermen—usually they have had no formal education and the availability of other types of jobs is very restricted in the area. If they did not benefit from the ease of fishing proffered by the lagoon area, most would probably have become marginal inhabitants of a metropolitan area, since they do not have financial opportunities to purchase advanced equipment for fishing in the open sea.

Total Value of the Ecosystem

Estimating the total value of ecosystems involves complex problems related to system boundaries, scale, thresholds, and aggregation of component values (Perrings 1995).

The total economic value (TEV) can be defined as the extent to which society would be worse off were the ecosystem to be lost, expressed in monetary terms. TEV (adapted from Pearce and Moran 1994) is calculated as follows:

$$\text{TEV} = \text{direct-use value} + \text{indirect-use value} + \text{option value} + \text{existence value.}$$

TABLE 5.5 INFORMATION ABOUT FISHERY ACTIVITY IN CANANÉIA

County	Fishermen Interviewed (N)	People per Family (ave.)	Fishermen per Family (%)	Family Members with Other Activity (%)	Fishing Site[a]		Fish Marketing Locale		
					Mangrove	Open sea	CEAGESP	Private	Union
Centro	108	7	22	6	75%	50%	87%	37%	6%
Carijo	127	5	2	3	87%	25%	62%	25%	12%
Acarau	20	7	25	0	33%	67%	100%	—	—
Morro	48	5	29	33	67%	55%	55%	33%	11%
P. Cubatão	88	6	21	7	93%	7%	7%	93%	—

[a]Percentage of people who fish in these areas.
CEAGESP, São Paulo State Supply Company.

TABLE 5.6 RESULTS FROM ECONOMIC VALUATION

Approach	*Function/Service Evaluated*	*Value ($/ha/yr)*
WTP	Existence value	230–710
TCM	Recreation	3,583
VCF	Fishery production (commercial)	470
HPA	Fishery production (artisanal)	228
	Total	$4,751/ha/year

The TEV values from the case study of the Cananéia mangroves are presented in table 5.6. It is important to emphasize that applying two distinct methodologies (CVM and TCM) to the same population does not necessarily imply a double counting, since each technique measured a different value. If we sum all the values in table 5.6, we will have an approximate value of the ecosystem without taking into account other important services that it could provide, as previously described under the indirect valuation approach.

Therefore, given the inevitable uncertainty involved in valuation processes, the precautionary approach based on maximal ecosystem conservation remains a high policy priority. As Perrings (1995) put it, a healthy ecosystem is valuable well beyond the benefits that may be computed in terms of consumptive preferences.

NOTES

1. For all proposed models, we are assuming a logistic distribution unless our results show otherwise. The logistic distribution is similar to the normal except in the tails, which are considerably heavier. The distribution that resembles it more closely is that of the t distribution with seven degrees of freedom. Therefore, for intermediate values of $b'c$ (say, between -1.2 and $+1.2$), the two distributions tend to give similar probabilities. The logistic distribution tends to give larger probabilities to $y = 0$ when $b'c$ is extremely small (and smaller probabilities to $y = 0$ when $b'c$ is very large) than the normal distribution. We should expect different predictions from the two models if our sample contains (1) very few responses (y's equal to 1) or very few nonresponses (y's equal to 0), and (2) very wide variation in an important independent variable, particularly if (1) is also true (Greene 1993).

2. The area used for the aggregated value was approximately half the total area of mangrove existent in the Complexo Estuarino-Lagunar de Cananéia-Iguape. The same area was used for calculations in the travel cost and household production approaches.
3. Standard error = 0.109; r^2 = 0.651; adj. r^2 = 0.632; Durbin-Watson = 1.2.
4. We know that this value is underestimated since statistics of the fishery sector in Brazil have been shown to be biased because of the difficulties in catch control.

6

Contingent Valuation in Brazil: An Estimation of Pollution Damage in the Pantanal

DOMINIC MORAN and ANDRÉ STEFFENS MORAES

The contingent valuation (CV) method is now acknowledged to be an appropriate approach for the valuation of nonmarket costs and benefits associated with a range of public goods including the environment (Mitchell and Carson 1989). Increased use world wide has been accompanied by both optimism and controversy. In particular, the role of CV in natural resource damage assessment in the United States has polarized opinions over the validity of hypothetical methods, and over the role of CV in resource allocation.[1] The number of applications in Brazil is now growing, but these survey mainly the willingness to pay (WTP) for amenity provisions such as water and sewage. This chapter details the more complex case of the valuation of pollution damage in the Pantanal.

Contingent valuation is a survey technique that aims to circumvent the difficult problem of valuing environmental resources by actually asking a sample of respondents about the value they place on some alteration in its supply or condition. The technique comes into its own whenever there is no alternative method to determine how much such nonmarket goods are actually worth in terms of some alternatively revealed willingness to pay. Thus the decision to take a trip to a particular beauty spot reveals—in terms of the trip cost—something about how the visitor values the site or some subset of its characteristics. In many cases, however, no such observation can be recorded and the use of CV is necessary to gauge how individuals value the resource, either as occasional visitors or even if they reside at some distance from the site and have never actually visited it.

The adoption of CV has much to do with the plurality of inseparable values associated with environmental goods, and with the plurality of motives held by resource users and nonusers (Cummings and Harrison 1995). As stated, revealed value studies, such as the travel cost method, offer the only plausible alternative validity check on hypothetical values, but impacts of interest are often difficult to capture in a recreational demand model. In contrast, the CV design protocol offers considerable flexibility for placing respondents in unfamiliar markets and for eliciting the *total* resource value. The limiting factors in this process are essentially cognitive.

It would be impossible to do justice to the many legitimate concerns that have arisen in this debate, and this paper addresses a relatively rudimentary set of issues related to the valuation of a complex ecosystem[2] such as the Pantanal. First, in the context of CV best practice, how easy is it to convey information on environmental alteration to elicit meaningful stated values? Second, what validity checks can be applied to prima facie satisfactory responses, and, by extension, what caveats should be attached to the results of CV surveys of complex goods? The study raises additional questions related to the commonly adopted approaches to analyzing and presentation of CV results. In contrast to existing CV studies conducted in Brazil (e.g., Briscoe et al. 1990), the study focuses on nonuse values related to ecosystem health and water pollution. The study is similar in spirit to recent applications to other complex goods, such as aquatic habitat (see Whittington et al. 1994), ground water cleanup (McLelland et al. 1992), valuation of landscapes (Willis et al. 1995), acid rain (Macmillan et al. 1996), and mangrove ecosystems (chapter 5).

SURVEY DESIGN[3]

Features common in the application of CV are (1) an introductory section that sets the general context for the decision to be made, (2) a detailed description of the good to be offered to the respondent, (3) the institutional setting in which the good will be provided, (4) the manner in which the good will be paid for, (5) a method by which the survey elicits the respondent's preferences about the good, (6) debriefing questions about why the respondent answered certain questions in the way he or she did, and (7) collection of individual respondent characteristics to statistically validate responses.

The protracted legal and academic dispute in the wake of the Exxon Valdez oil disaster has culminated in a set of recommended guidelines for conducting CV studies that are selectively adopted in this study (NOAA 1993).[4] One of the most contentious recommendations relates to the use of a referendum-based discrete-choice (DC) format, rather than an open-ended format, for asking WTP questions. Unlike the open-ended WTP question, which asks for a free statement of an amount, this variant asks respondents simply to state whether or not they would vote yes to a program to pay a pre-set amount of money (usually in tax) to avoid the suggested decline in environmental quality.

DISCRETE-CHOICE MODELING PROCEDURE

Despite its alleged incentive compatibility, the DC format introduces numerous unresolved design problems related to choosing the bids offered to respondents, and to the choice of appropriate models (Cooper 1993; Alberini 1995). The format also offers a potential diagnostic approach for checking response cognition. This chapter concentrates on the development of the format in the context of the Pantanal case study.

The theoretical approach to modeling qualitative choice data is well known (Johansson 1987; Greene 1993). Alternative *parametric* approaches are based on either the random utility model (RUM) (Hanemann 1984), or the cost function approach (Cameron 1988). McConnell (1990) shows that in certain circumstances the approaches are exactly dual to each other. In essence, both approaches attempt to approximate the expected value of a random WTP variable (in this case, the pattern of discrete WTP responses):

$$E(\text{WTP}) = \int_0^\infty 1 - F(b)\} \, db \qquad (6.1)$$

using a distributional assumption for $F(b)$ to represent the stochastic element of the RUM, and which can be modeled using standard binary choice methods. Logit or probit models [employing the exponential (logit) or normal (probit) cumulative distributions] are typically chosen to link the probability of an event taking place (in this case, a yes response to "WTP R$?") to the selected utility difference (ΔV), known as the index function,

$$P_{\text{yes}} = (1 + e^{-\Delta V})^{-1}. \qquad (6.2)$$

There is nothing in economic theory to say how WTP should be distributed, and parametric approaches can equally use alternative distributional forms such as the Weibull, lognormal, and gamma, which nest common forms for certain parameter values[5] (see Lee 1992 for the relative properties of these distributions). The distributional assumption and the utility difference (ΔV) are the two model choices in parametric analysis. It is also possible to dispense with the assumptions underlying both parts of the model and to use *nonparametric* methods (Kristrom 1990).

Hanemann (1984) shows that a linear form for ΔV results from the difference between two underlying utility, and theoretic indirect utility, functions, giving the model

$$P_y = \frac{1}{1 + e^{-\alpha+\beta A}}, \quad (6.3)$$

where A is the *bid* variable. An alternative, commonly adopted log-logistic form—including the log of bid A—can be shown to avoid predictions of negative WTP, but it frequently introduces problems with tail values for certain parameter values.[6] Hanemann and Kanninen (1996) discuss the pros and cons of model development.

Bid Design and Deriving Mean/Median WTP

Unlike the open-ended variant, which simply leads to an arithmetic mean, the DC mean WTP is derived from the parameters of the model used to predict the function bounding the expected E(WTP) (see Hanemann 1984). Alternatively, it is possible to integrate over the area bounded by the function (Loomis 1988). The median can be inferred where the function crosses probability yes = .5. Either way, it is important that the predicted function be well defined in the sense of being closed at or near the highest value offered in the discrete-choice survey.

The trick in getting this function to be well behaved lies partly in the design of the amounts offered and partly on the distributional assumption made. The expectation described previously assumes that $F(b)$ has a lower limit at zero (i.e., that nobody will say no at WTP \$0), and an upper limit at 1 as bid amounts tend to infinity (that is there is some bid amount high enough to induce a certain negative response). Graphically this implies a sigmoid 's' function in the probability space requiring a range of bids offered (combined with an appropriate distribution) to ensure that the extremes of the predicted function are 'banged down.' This amounts to the unusual econometric task of *designing* one of the explanatory variables of the regression for the desired response function, to ensure model parameters are unbiased and minimum variance. A considerable literature on DC design shows the difficulty of optimally locating bids prior to knowing the actual pattern of responses to the amounts offered (Kanninen 1995). This means that the investigator must use judgment and any information from an open-ended pretest. The problems with the DC method tend to be associated with inappropriate bid designs leading to fat tails where the function remains open at the highest bid (see Randall and Kriesel 1990), or too short a bid range leading to insufficient response patterns (all yes or all no). In the former case, it is typically necessary to truncate the function to avoid the predication of very high mean values.[7] In the latter case, the response function may be flat over the bid range, leading to infinite WTP values or, sometimes, a function that does not cross .5.

There is currently some disagreement as to how to circumvent the problem, although for common distributions like the exponential, rules of thumb can be derived from analogous problems in biostatistics (see Kanninen 1995; Hanemann and Kanninen 1996). Typically five- to ten-bid values are chosen based on the range of WTP values revealed by an open-ended pre-test survey. Other design criteria are possible (e.g., Jordan and Elnagheeb 1994).

THE PANTANAL

The unrivaled wildlife habitat of the Pantanal offers a particularly stringent challenge for contingent valuation. The area is subject to several forms of extractive (subsistence) and nonextractive uses (tourism) from which market-based valuation information may be derived. The passive-damage nature of several current environmental threats suggests a role for CV to explore aspects of total value to be considered in potential regulation of agricultural and mining activity.

Centro de Pesquisa Agropecuária do Pantanal (EMBRAPA-CPAP) identified a number of threats to the Pantanal environment, three of which form the basis of a damage scenario presented to survey respondents:

a. Mercury pollution from informal gold and mineral mining
b. Agricultural run-off and sedimentation resulting from land use change in the adjacent plateau
c. Agrochemical residues.

In addition, the potential threat posed by planned engineering investments on the Paraguay-Paraná waterway project known as Hidrovia provides an interesting backdrop to the study (see Bucher and Huszar 1995). Although the nature and scale of any potential damage remain uncertain, this study undertakes to investigate the role of CV in any prospective social cost-benefit analysis.

Scenario

Elements one through four of the design process (see preceding section, Survey Design) comprise the method of scenario communication. A review of much of the CV literature shows scenario communication to be a particularly weak area in reducing complex and emotive issues to the dimensions of heuristic devices such as photographs, pie charts, and quality ladders. Use of such methods in risk communication has been shown to bias resulting welfare measures (Loomis and duVair 1993), and even extensive pre-tests cannot guarantee cognitive validity. The two main issues are (1) whether the respondents' interpretation of the scenario corresponds to that intended by the researcher, and (2) whether or not the scenario actually corresponds to the eventual outcome of a particular project. In general, the design of a scenario for goods with which respondents have direct-use experience will pose fewer problems than the nonuse case.

Given the three main causes of damage, the problem is what to ask about. Photographic evidence was not considered sufficiently subtle to convey an idea of ecosystem damage. The preferred scenario read to respondents informed them of the damaging effects of a decline in water quality in the Pantanal, and of its direct influence on plant and animal diversity. Damage was attributed to the three factors (a through c) previously listed, with additional effects related to sporadic deforestation and civil constructions, such as roads, minor dams, and dikes. Some of the complexity of conveying damages to respondents was avoided by using a species box showcard (suggested by Macmillan et al. 1996). The scenario card (see the appendix to this chapter) suggested three ecosystem damage stages based on current patterns of resource use in the Pantanal, and it placed respondents before an impending reduction from stage B to stage C over the period from "today" to the year 2010. A further map showcard (see figure 6.4) informed respondents that the extent of damage might be locally severe in specific areas of the Pantanal. The scenario continued to inform respondents of the costly nature of new and additional control technologies over and above current payments made by all taxpaying Brazilians, as well as through current use charges for individual anglers, which are paid in two ways: annual licence and lacre (seal). These contributions notwithstanding, controlling water pollution in certain areas such as the Pantanal was described as requiring additional expenditure. One important consideration in the determination of fund allocation was the value individuals placed on the characteristics of the Pantanal environment and the leisure opportunities it afforded, hence the reason to elicit individual willingness to pay for the maintenance of current water quality conditions.

As a prelude to an initial filter question on the willingness to pay (anything), several standard qualifying statements reminded respondents of the purpose of any payment, the likely outcome if sufficient funds were not forthcoming, their budget constraints, current and future uses, and the availability of substitutes in Mato Grosso.

Question Format and Administration

Points 5 through 7 (in the preceding section, Survey Design) relate to the choice of surplus measure, the preferred payment vehicle, and the statistical analysis of data for the calculation of a mean or median surplus measure.

The choice between the willingness to pay (to prevent a quality decline) and the willingness to accept compensation (for a decline that has actually happened), depends on the property right over the good (Mitchell and Carson 1989). As in most recent examples, a WTP question format[8] was used corresponding to the equivalent surplus of a potential quality decline. The format makes an implicit assumption regarding the property right, which was not contested by respondents.

The choice of an appropriate payment vehicle has been found to have a significant effect on WTP (Bateman et al. 1995). The Brazilian experience with CV has been limited to one study on rural water (Briscoe et al. 1990), which used a hypothetical monthly tariff for connection. Forms of payment for less familiar goods are controversial everywhere, although initial doubts that such scepticism would add to considerable suspicion of authority and distrust of government were largely unwarranted. A tax option offers considerable advantages in being an egalitarian means of payment, but it is often rejected as unpopular.

Alternative use-related vehicles include an annual licence payable by all recreational anglers at a cost of R$34, or the previously mentioned lacre, which is fixed on catch boxes at a cost of R$4 per box[9] (volume). Neither mechanism is wholly ideal as a payment vehicle. The liability of the lacre is clearly dependent on catch volumes and some respondents may have ruled out future liabilities on this basis. On the other hand, there is a definite feeling that the catch-related payment offers respondents a closer approximation to a payment for the environment, and this helps to circumvent scepticism over a wasted payment. Some respondents feel that the higher-priced licence translates more clearly into an annual payment. The obligation to pay anything extra is nevertheless conditional on the decision to renew.

This uncertainty dictated a split-sample approach in an open-ended[10] pre-test using both lacre and licence to discover the bounds for the subsequent discrete-choice experiment.

The preliminary questionnaire consisted of thirty questions accompanied by eight showcards offering multiple-choice answers and providing supplementary information to a verbal scenario. Several variables, such as fish catch, repeat or first visit, group or family size, and days spent in the area, may also serve as validatory covariates in bid functions.

Questionnaire versions were designed to be administered to a sample frame of tourists and recreational anglers passing through the towns of Corumbá and Miranda (MS). Four trained interviewers were employed in Corumbá and another two in Miranda over a period of four months between August and November 1994. Interviews of approximately twenty to twenty-five minutes were on the basis of casual intercept while respondents had catches weighed by the Forest Police.[11] A total sample of 586 visitors of a total recorded annual population of 110,000[12] were interviewed at Forest Police stations. Willingness to participate was ascertained by casual intercept using a method that has been employed in conjunction with both interview and self-administered CV studies (see Boyle et al. 1994). Despite the need to maximize interview numbers (and a low visitor throughput at selected sites), interviewers were instructed to interview group members simultaneously, and to approach male and female visitors equally.

Sample Frame

Determination of a sample frame from an affected population presented a problem for this study (and potentially a problem for many nonuse CV studies in developing countries). A sample frame bias arises where "the sample frame does not give every member of the population chosen a known and positive probability of being included in the sample" (Mitchell and Carson 1989:261). The problem jeopardizes the accuracy of projecting the results of this survey to any population beyond the socioeconomic group comprising the majority of respondents who themselves will be a subset of potential interviewees.

While this study is dealing with a unique and world-renowned resource (i.e., with a high nonuse component that potentially widens the population to the whole world), the potentially affected population was in the first instance determined to be made up of direct users of the resource. In other CV studies, this is typically a homogeneous group, to the extent that an unmodified instrument is considered appropriate for all prospective respondents.

In this study, the population of users was determined to span a range of socioeconomic classes and income groups that could not reasonably be bridged by a single survey instrument. In particular, it was felt that an unmodified survey asking poorer groups about water quality in relation to biological diversity would probably result in a high protest rate. On the other hand, use of multiple survey versions raised the prospect of having to adopt specific communication devices and payment vehicles, with no guarantee that the subject good is perceived equally across versions.

The issue of how to sample from a highly heterogeneous user population therefore presents a problem for CV surveys. A single survey instrument rules out probability sampling (as advised by NOAA), the objective of which is to obtain a sample similar to the parent population. We deliberately restrict the population to a subset of anglers, and the resulting selection bias is accounted for when aggregating over the relevant population.

Model Estimation

This section summarizes parametric results from the single DC question, plus the interval model of the double-bounded format. The choice between the models essentially comes down to professional judgement while favoring conservative estimation procedures.

The design method here is Cooper's optimal equal area bid design method for minimizing the mean square error of the resulting WTP estimator (Cooper 1993).[13] The two-stage procedure of design algorithm requires a distributional assumption for the preferred open-ended data, to set the parameters on a bid selection process. Cooper's design places considerable weight on getting the "right" open-ended sample for the design, recommending several preliminary samples to assess response consistency and the possibility of diffuse information in a small sample. As this is not always an option, some discretion is necessary when following the design procedure which in most circumstances will not be "optimal."[14]

Pre-test data from the lacre and payment card (see appendix) gen-

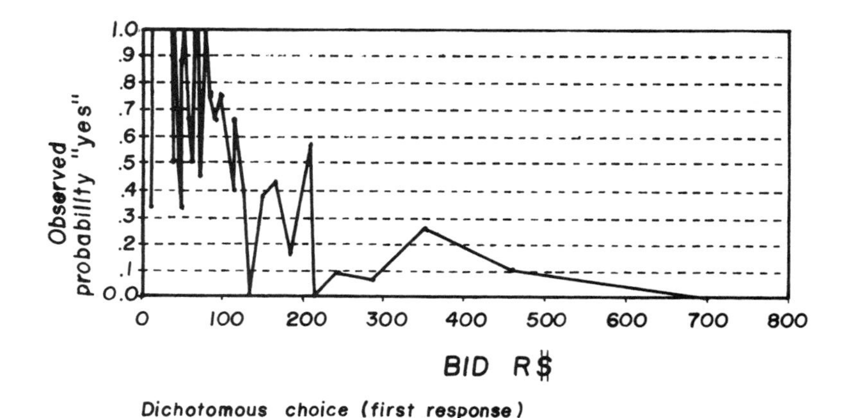

FIGURE 6.1 Plotted Proportions of Yes Response, by Value of Bid, in R$ (Brazilian Reals)

erated the most conservative pre-test bid range. However, the lacre does not imply the same annual commitment implicit in the purchase of a licence. A compromise was to use the conservative bid range to set the bid amounts, but to use the licence as the payment vehicle in the DC questionnaire.

A Box-Cox test revealed the pre-test data to be lognormally distributed. A lognormal bid vector was designed for the administration of 400 surveys, allowing the DWEABS (distribution-weighted equal area bid system) algorithm to design an optimal vector without specifying a unique number of bids. The resulting (fairly untypical) vector involved fifty-six bid values (ranging between R$1 and R$701), to be administered to subsample sizes ranging from one to fifty-two respondents. Given recent findings favoring the efficiency of double-bounded models (Hanemann and Kanninen 1996), these values were halved and doubled.

Logistic Regression

Regardless of the selected distribution of the bid vector, it seems impossible to speculate a priori about the best-fitting model of actual observed responses. Although most studies automatically adopt the logit model, the best procedure should be to vary the distributional form and compare models on the basis of predictive ability. The plotted proportions for the single DC response (figure 6.1) provide an additional clue about behavior of tail. The lognormal may be dis-

TABLE 6.1 LOGIT MODEL[a] (First Bid)

Variable	*Coefficient*	*t-stat.*
Constant	1.7663	6.975
Bid	−0.01113	−7.709

[a]Logit regression (1/0 dependent variable).
$n = 267$.
Log-likelihood = −123.83.
Restricted log-likelihood (slopes = 0) = −184.39.
McFadden's $R^2 = 0.33$.
% correct predictions = 79%.
$y = 1 \rightarrow 124$; $\psi = 0 \rightarrow 143$.

counted on intuitive grounds because of its fat right tail, which is also an undesirable characteristic of the log-logistic model.

A common linear logistic (equation 6.3) was selected to analyze the single-response DC questions, based on the idea that the incentive-compatible properties of the format should best be reflected in the single question. The logistic version will be the only one generalized to evaluate the effect of additional covariates.

The coefficients of the univariate linear logistic estimated by maximum likelihood are shown in table 6.1. Accordingly, for the linear form, this restricted model correctly predicts 79 percent of the yes and no responses actually observed, with the *bid* variable highly significant. Note that in contrast to the ordinary least square (OLS),[15] the omission of other relevant explanatory variables has the effect of biasing the coefficient b towards zero (for a proof, see Cramer 1991:37). As such, it is necessary to first compare the performance of the model with additional covariates to be as sure as possible to avoid any bias in mean/median WTP. Several matters are open to debate, particularly whether to include the income variable, and how to reflect the marginal utility of money between two utility states. Omitting income is tantamount to assuming constant marginal utility of income between states under consideration. Desvouges et al. (1992:67) point out that this can be the case only for perfectly replaceable commodities. Including income allows for a test of perceived uniqueness, in the sense that varying the marginal utility of income with the level of environmental change (implied by a significant coefficient) would not occur if perfect substitutes were available.

Initially, five additional variables were codified to be included in the regression analysis. Table 6.2 presents the maximum-likelihood

TABLE 6.2 MULTIVARIATE LOGIT MODEL[a]

Variable	*Coefficient*	*t-stat.*	*Definition and Coding*
Constant	−0.56403	−0.54874	—
Bid	−0.01195	−7.46889	WTP discrete amount R$ (Q. 22)
Income	0.00023	3.40298	Selected household annual income band (all sources) (Q. 27)
Visit	−0.01664	−0.67697	Total number of visits made to the Pantanal area (Q. 6)
Know	1.26476	3.61938	Prior knowledge of the pollution problem: 1 = yes. 0 = no (Q. 20)
Catch	−0.00413	−1.28193	Total fish catch by party (all species) in kg (Q. 19)
Age	0.02399	1.04733	Respondent age (yr) (Q. 26)

[a]Additional covariates in logit regression.
n = 267.
Log-likelihood = −100.26923.
Restricted log-likelihood (slopes = 0) = −184.39.
McFadden's R^2 = 0.40.
% correct predictions = 83%.
χ^2 (6) = 148.25 (critical value = 12.59).
Q., question (see appendix).

parameter estimates of the same variables and related diagnostic statistics. Variables *bid, income,* and *know*[16] are significant at less than the 1 percent level and have the expected signs. In the latter case, it is expected that informed respondents are more disposed than those learning of the problems for the first time. This is a potentially interesting finding as the knowledge issue remains a sticking point in CV deliberations. Specifically, can people value things they know nothing about? Variables *catch* and *age* (significant only at 0.19 and 0.29 levels, respectively) provide somewhat counterintuitive results. In the case of *age,* the positive coefficient suggests that WTP increases with age—typically the reverse is true—while WTP seems to be inversely related to the variable *catch.* A plausible explanation for the latter relates to the interpretation of the last variable *visit,* which is the least significant variable of the regression, although of negative sign. Either of these variables is the quantity variable in a demand relationship. Both are in the range $-1 < d < 0$, which is consistent with economic theory and implies that the hicksian demand curve will be downward sloping.

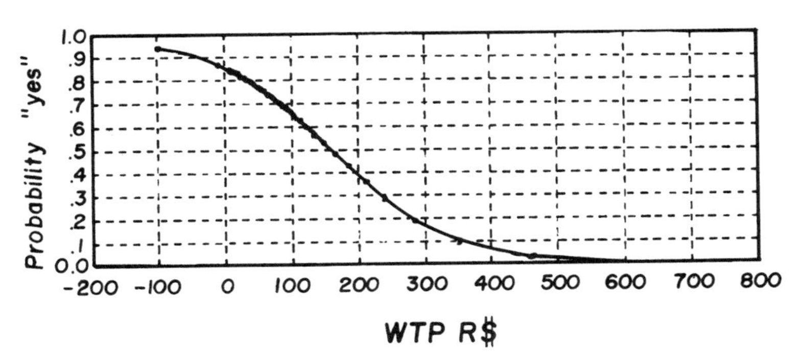

FIGURE 6.2 Parametric Analysis of Predicted Logit Function for Dichotomous Choice Variable (Proportion Responding Yes) by Value of Bid, in R$

Mean/Median Estimation

Plugging the estimated coefficients from table 6.1 into the logit model, we have

$$\text{Probability (odds) “yes”} = \frac{1}{1 + e^{-[1.7663-0.01113(\$A)]}}, \tag{6.4}$$

which is the predicted function in figure 6.2. Note, however, that extending the plotted function shows high probabilities of negative WTP (e.g., the probability of a yes response to willingness to accept \$100 is about .95). Reading off from probability yes = .5 shows the median lying around 150 real. This can be validated by solving equation 6.4 for Pr (yes) = .5, as follows:

$$\log(\text{odds})\frac{0.5}{1 - 0.5} = 1.7663 - 0.01113(\$A), \tag{6.5}$$

giving

$$\$A = \frac{1.7663}{0.0113} = 158.69, \tag{6.6}$$

which is the formula α/β for the unrestricted mean[17] including the positive and negative parts of the predicted function (see Hanemann 1984). Since the survey rules out willingness-to-accept statements, the appropriate restricted mean is

TABLE 6.3 DOUBLE-BOUNDED MODEL

Variable	*Coefficient*	*t-stat.*
Constant	0.61334	2.102
Bid	−0.01296	−11.04
Income	0.00020	4.213

$n = 265$.
Response cases: yes-yes = 78; yes-no = 44; no-yes = 36; no-no = 107.
Log-likelihood = −264.832.

$$E^{+}(\text{WTP}) = \frac{\ln(1 + e^{1.7663})}{0.0113}, \tag{6.7}$$

which truncates negative values, giving a higher mean of 172.92. Unless otherwise stated, and consistent with the view of the conservation of the Pantanal providing positive utility, the restricted model shall be used.

Double-Bounded Estimation and Preference Uncertainty

A recent modification to the DC format is the addition of a follow-up question in which respondent's responding yes (no) to an initial DC question are offered a subsequent WTP amount higher (lower) than the first bid . Respondent willingness to pay is therefore identified by "tighter" interval-censoring combinations of yes yes, yes no, no yes, and no no, which gives more information on the underlying willingness to pay and can be statistically more efficient (Hanemann et al. 1991; Kanninen 1995).

Table 6.3 contains the parameter estimates of a model used to calculate the mean for comparison with the single response. These parameters produce a response probability function similar to that of the single response, apart from an extended right tail, now predicting that the probability of observing a yes to the value of R$701 is reduced to 0.000862 (from a value of 0.002391 shown by the response function of the single response), while that for the highest possible bid 1402 is only 7×10^{-7}. Evidently, the double-bounded model is giving less weight to amounts that were not accepted, than the single response model, while the integral of the function beyond the original highest bid adds a negligible amount to the density. This is precisely because the latter provides an opportunity to retrieve some of the respondents who refused the original highest bids, allowing more information of the shape of the distribution somewhat nearer the true sample moments.

TABLE 6.4 LOGIT MODEL[a] (Second Bid Only) and Estimated Means

Variable	*Coefficient*	*t-stat.*
Constant	0.55396	2.99229
Bid	−0.00098	−1.21249

[a]Logit regression (1/0 dependent variable).
Restricted mean point estimate = 1028.7 (465.84–20081.74).
$n = 265$.
Log-likelihood = −178.00.
Restricted log-likelihood (slope = 0) = −178.74.
McFadden's $R^2 = 0.004$.
% correct predictions = 56%.
χ^2 = test for regression significance = 1.48 (significant only at 0.22).
$y = 1 \rightarrow 158$; $\psi = 0 \rightarrow 107$.

Value Uncertainty

Use of multiple response CV formats raises the interesting issue of the stability of respondent preferences between questions. Extreme revisions of WTP statements may point to unstable preferences and, by extension, the need to reappraise the assumptions made by both respondent and investigator in evaluating elements of the scenario. A possible test of stability is to analyze first and second responses separately, and to conduct some type of nonparametric distance test for two distributions. Cameron and Quiggin (1994) have suggested the use of bivariate models to empirically test the hypothesis of the similarity between first and second responses. However, Aberini (1995) has questioned the power of the test of any divergence between responses, and he recommends the use of the regular interval model proposed by Hanemann et al. (1991).

For present purposes, it is sufficient to analyze the second responses of the sample in isolation from the first response[18] to investigate the cause of any likely divergence between responses.

Table 6.4 repeats the single response logit estimation, this time for the second response. The consequences are somewhat alarming. First, the regression is significant only at the 22 percent level, predicting 56 percent of responses correctly. A plot of the response proportions shows that the erratic acceptance (rejection) of several very high (low) follow-up values leads to a predicted logit function (for the second responses) that is almost flat, such that the extrapolated tails are likely to be fat for very high positive and negative values. The result is that integrating for the restricted mean results in a very high point estimate of 1028.7 (465.8 to 20081.74), whereas an unrestricted mean of 565.33 is not significantly different from zero[19] at the 95 percent level of confidence. Similarly, in contrast to the first response, the median is located in the vicinity of R$550. Clearly, these results in their own right would be unacceptable.

It is of some interest to speculate about the behavior underlying this observation, and there are several possible explanations. The first relates to latent yea-saying behavior on the part of respondents, which is exacerbated by a follow-up question. Clearly, the plotted proportions suggest that some very high values have been accepted as follow-up bids, resulting in insufficient response variation relative to the "typical" function associated with the first response. Yea-saying behavior has been popularly ascribed to forms of symbolic commitment or social desirability on the part of respondents (Holmes and Kramer 1995). The yea-saying hypothesis contradicts the theory of anchoring behavior, which has been used to explain the lower mean value typically derived using a double-bounded model. According to this view, respondents systematically reject any follow-up values once they are told an initial value that they consider to be the "true" value.

A much cruder suggestion is that, far from helping to narrow the range of respondent WTP, the follow-up question merely provokes an honest response, reflecting the individual's complete ignorance of her value for the subject good. Such a reaction may provoke yea-saying and will be detrimental to model estimation irrespective of the bid selection method. However, the bid selection procedure does offer a related explanation of the lower mean estimate. In this case, the skewed bid positioning dictated by the area of the lognormal bid distribution concentrates many low values close together in the main part of the density, and relatively few observations of high bid values in the tails of the density. The universal acceptance of low values (and their follow-up) does not disturb the predicted function and provides more information from people who initially reject the initial bids. The same is true for high bid values, although unexpected acceptances are sufficient to upset the behavior of the whole function. This problem demonstrates how the variance of the function parameters can be rescued by increasing sample size at all bid amounts.

The result found here is likely to be caused by a combination of yea-saying behavior and poor design. The perils of poor bid distribution and an ill-defined good are evident.

TABLE 6.5 MEAN ESTIMATES

Model	*Mean WTP*[a] *(R$/yr.)*	*Conservative Mean*[b]
Open-ended	52.76[c] [35.09–70.39[e]]	—
	89.74[d] [74.2–103.8[e]]	—
Single-bounded	168.29 [144.34–200.15[e]]	137.24 [119.43–159.66[e]]
Double-bounded	137.51 [121.71–156.15[e]]	112.93 [98.70–128.32[e]]
Bivariate probit	215.25 [196.37–235.80[e]]	—
Nonparametric single-bound (Kaplan-Meier)	346.10 [315.00–376.86[e]]	—

[a] $n = 265$.
[b] $n = 364$.
[c] Lacre payment vehicle and payment card.
[d] Licence payment vehicle, no payment card.
[e] 95% confidence intervals by Krinsky and Robb method.

Comparison of Means and Aggregation

Table 6.5 presents mean estimates from the cleaned data set for the models described, plus results from a nonparametric alternative that dispenses with the distributional assumption and index model (see Kristrom 1990). Estimation uses a regular and "conservative" data set[20] with confidence intervals generated using the method of Krinsky and Robb (Park et al. 1991). The interpretation given to the results derived from the licence questionnaire is of a mean annual WTP *in addition* to the current licence price. All DC means are at least inside the bid vector from which they were obtained,[21] although the variability between DC means and their size relative to the open-ended estimates is disconcerting.

This study adds to the growing list of studies finding a disparity between open-ended and dichotomous-choice welfare estimates (Kristrom 1990). However, in making such a comparison, it is impor-

TABLE 6.6 AGGREGATIONS SCENARIOS

Pantanal of Mato Grosso do Sul (110,000 visitors p.a)[a]		
Mean	*Aggregate value (R$)*	*95% confidence interval*
Open-ended	5,803,600	3,859,900–7,742,900
Double-bounded DC	15,126,100	13,388,100–17,176,500

[a]For lower bound figure for all Pantanal multiply by 2.

tant to note that a particular distributional assumption is made for the discrete-choice responses that does not apply to the open-ended data. In drawing conclusions, the extent to which inappropriate models and distributional assumptions are driving the disparity is uncertain.

Because of the advantages offered by the double-bounded format relative to the single response, the double-bounded mean is the favored estimate from the DC format. This mean is nevertheless some 2.6 times the most conservative open-ended mean. Given the need for contingent valuation studies to err on the conservative side, the promotion of the referendum format by the NOAA panel seems somewhat questionable, particularly if a central tenet of the method is that respondents know their own preferences.

For the purpose of aggregation (table 6.6), the extent of the market is already predetermined by the decision to restrict the survey to a sample of visiting anglers. This decision essentially violates the condition for probability sampling (giving each individual a known probability of being selected) and has inherent limitations for aggregation and prediction. In the former case, this means that the appropriate population for aggregation is the number of anglers registering fish catches in any one year in the Pantanal. This choice underestimates social value, particularly when the resource has a high nonuse value and high subsistence use.

Another drawback of the restricted sample frame is that the model is essentially calibrated with parameters drawn from a population that does not represent the majority of resource users. In other words, the model cannot be used to predict the WTP for individuals other than those in the sample frame. When the latter is derived from optimal probability sampling, aggregation is not a problem. The problem occurs in the event that resource use is characterized by heterogeneous socioeconomic groups who cannot all be sampled using the same survey instrument. A compromise involves either using the preferences of one high-profile subsample of users, or designing alternative questionnaires (with the pitfalls this might entail when reconciling information provision and cognition). This problem is likely to be accentuated in developing countries and has not received much attention in the literature.

CONCLUSION

As is typical in many CV papers, the conclusions of this study cannot be wholly dismissive of the CV approach, providing one subscribes to the sovereignty of individual preferences. Accordingly, from a policy perspective, these results show that there is a considerable nonmarket value attached to the Pantanal environment.

One or two caveats should be stressed. First, the obvious cognitive limitations on the object being valued should be repeated. Was the scenario sufficient, and what exactly were respondents valuing? Methods to validate the cognitive effects of heuristic surveys are vital. That the construct of the investigator is the one received by the respondent, is an article of faith that cannot be checked. This is precisely what makes CV at once so vulnerable and yet unassailable. Some psychological input (i.e., focus groups, postsurvey debriefing, etc.), to the design process may reveal where potential problems lie.

Several problems associated with the use of the DC format remain unresolved. Nevertheless, modeling procedure itself can also accommodate expressed uncertainty in responses (see, for example, Li and Mattsson 1995). The issue of uncertainty is particularly germane to the valuation of species and ecosystems. It is not inconceivable that such analysis would present some rigorous challenges to the standard theories of consumer choice (Hanley et al. 1995). This study has followed CV protocol and seems to be consistent with as much as demand theory can say on price–quantity relationships. Using criteria proposed by Kanninen (1995) to assess bias and variance of the estimator, the model performs well despite the unorthodox bid vector. The resulting mean estimates might reliably be proposed.

Finally, it is important to stress that the sample frame of this study is limited, and that this is reflected in the aggregation procedure. The prevailing income distribution and patterns of use of environmental resources in Brazil present particular problems for the use of CVM—in particular, there is a need for designs capable of bridging the marked socioeconomic differences between stakeholders. A more plural approach to environmental valuation may well reveal unbridgeable perceptions of environmental change.

APPENDIX

SURVEY QUESTIONNAIRE

PANTANAL ENVIRONMENTAL CHANGE SURVEY—CONFIDENTIAL

SERIAL NUMBER: ☐☐☐☐-☐☐

INTERVIEW NUMBER: ☐☐☐-☐☐☐

DATE: ☐☐ 1994

Time interview started (24-hour clock): _____ : _____ hour

Researcher's use only:
____ Complete
____ Incomplete
____ No payment
____ Protest

GENERAL INSTRUCTIONS FOR THE INTERVIEWER

1. Interview only individuals, avoiding participation of others from the same group. (However, you may interview more than one person from each group, as individuals.)
2. Mark answers clearly. Note your personal comments when you have doubts.
3. Under normal circumstances, do not interview those under 18 years of age.
4. If interviewing a family group, you should try to interview the head of household.
5. Try to interview both men and women, and mark the sex:
   1. ___ Male 2. ___ Female
6. Read the interview questions aloud, skipping ahead when instructed.

THE QUESTIONNAIRE

Hello, I'm __________ (name) from EMBRAPA. We are carrying out a survey of people who use the Pantanal, and I'd be grateful if you would answer a few questions. The information that you provide will be kept **strictly confidential,** and it will be used only for statistical analysis. **I shall not be asking your name or home address.**

If answer is YES, then proceed. If answer is NO, then withdraw politely.

First, I would like to get some basic information regarding your visit.

1. Are you on holiday, are you working here, or do you live here?
   a. ___ Holiday (Go to Question 2)
   b. ___ Working in the region (Go to Question 6)
   c. ___ Live here (Go to Question 7C)
2. Is this your first visit to the Pantanal?
   a. ___ Yes (Go to Question 3)
   b. ___ No (Go to Question 6)
3. Will you visit again?
   a. ___ Yes (Go to Question 4)

b. ___ No (Go to Question 5)
c. ___ Don't know (Go to Question 7B)

4. How often do you think you will visit in the next 12 months?
_____times (Go to Question 7B)

5. Why will you not be visiting again?
_____ (Go to Question 7B)

6A. How many times have you visited the Pantanal?
_____ visits.

6B. How many times have you visited the Pantanal during last 12 months? (Include today's visit.)
_____ visits. (Go to Question 7A)

7A. How many days do you normally stay in the Pantanal?
_____days. (Go to Question 7C)

7B. How many days have you stayed in the Pantanal?
_____days.

7C. Have you to been to any of the following cities? (Read the list aloud and mark.)
a. ___ Coxim (MS)
b. ___ Taquari (MS)
c. ___ Poconé (MT)
d. ___ Cuiabá (MT)
e. ___ Nossa Senhora do Livramento (MT)

8. How many people are in your party today including yourself?
a. ___ 16 or over?
b. ___ Under 16?
c. ___ I'm alone

9. How many people are in your **immediate family household** who are **not** with you today?
a. ___ 16 or over?
b. ___ Under 16?

10. Where do you live? (Town/village and county only, **not home address**)
_____

11. How far away is that? (kilometers)
_____km

12. How do you travel to the Pantanal?
a. ___ By road (Go to Question 13)
b. ___ By air (Go to Question 16A)
c. ___ Live here (Go to Question 16A)

13. Did you come directly to the Pantanal from your town, or did you stop in other locations before arriving here? (For any reasons: walking, fishing, meeting friends or group, etc.)
a. ___ Came directly to the Pantanal (Go to Question 16A)
b. ___ Stayed in other places before arriving here (Go to Question 14)

14. What is the last town you stayed in before arriving here?
______

15. How far away is that? (kilometers)
______km (Go to Question 16B)

16A. Approximately how long did it take you to go from your residence to here?
______hours (Go to Question 17)

16B. Approximately how long did it take you to go from this last location to here?
______hours (Go to Question 17)

READ ALOUD THE FOLLOWING:

I would now like to ask you some more specific questions about what you value in the Pantanal and how much money you spent here.

17. This card (indicate showcard 1) lists a number of reasons for coming to the Pantanal. From the list, please select your **main** reason for coming here. Choose **one** only.

Showcard 1

a. ___ Possibility of catching big fish
b. ___ Possibility of catching a lot of fish of any size
c. ___ Possibility of catching different species of fish
d. ___ Proximity to and accessibility from where you live
e. ___ Proximity to other fishing regions
f. ___ Possibility of seeing wildlife
g. ___ Environmental quality (beautiful, unpolluted landscape)
h. ___ Other (please specify) ________________

18. This card (indicate showcard 2) lists possible expenditures **you and your fishing party** may have made on this fishing trip. I would like to know approximately how much money **you and your fishing party** spent on each item. (If you didn't have a particular expense, enter zero.)

Showcard 2

a. Petrol for the trip (by motorcar) R$ ______
b. Tackle and fishing equipment R$ ______
c. Air tickets (per person) R$ ______
d. Bait and ice R$ ______
e. Fishing guide services R$ ______
f. Boat and motor rental R$ ______
g. Petrol and oil for the boat R$ ______
h. Food and drinks R$ ______
i. One-time payment* for items (a) to (h) R$ ______
j. Other R$ ______

*Such a package is common in some regions of the Pantanal. The interviewers are instructed to circle the letters of the items included in the package.

INTERVIEWERS, BE AWARE:

1. If the respondent answers in quantities (petrol liters, amount of bait, etc.), note those quantities without attempting to calculate their monetary value.

2. If the respondent has difficulty estimating the costs of each item, ask for an estimation of the total *amount spent: R$* _____

19A. This card (indicate showcard 3) lists some fish species **you and your fishing party** may have caught on this fishing trip. Please list the total number of each species caught, and/or your best estimate of the total weight of each of these species. **Do not include purchased fish.**

Showcard 3

	Caught		*Total Number*	*Total Weight (kg)*
	Yes	*No*		
Pintado/cachara	___	___	_____	_____
Dourado	___	___	_____	_____
Jaú	___	___	_____	_____
Pacu	___	___	_____	_____
Curimbatá	___	___	_____	_____
Piranha	___	___	_____	_____
Tucunaré	___	___	_____	_____
Piraputanga	___	___	_____	_____
Barbado	___	___	_____	_____
Other	___	___	_____	_____

19B. Did you buy some fish to take away?

a. ___ Yes. How many kilograms? _____

b. ___ No

READ ALOUD THE FOLLOWING:

I would now like to tell you about some possible environmental changes in the Pantanal.

Environmental changes, mostly in the adjacent highlands, have a **negative influence** on the water quality in the rivers of the Pantanal. **Water quality directly affects the abundance of animals and plants** living in the water or using the water as a resource.

The main **causes** of environmental change in the Pantanal that originate in the adjacent plateau are the following:

1. Land use change in the adjacent highland resulting in agricultural run-off, **sediment,** and **agrochemical residues** carried into the Pantanal rivers.
2. Gold mining, causing **mercury pollution** and **silting** in Pantanal rivers.

Within the Pantanal, there are environmental changes too. These are caused by **deforestation** (for the purpose of introducing grasses) and **civil construction** (roads, minor dams, and dikes).

This card (indicate the scenario card) describes the **damage that has occurred** in some areas of the Pantanal as a result of these environmental changes, and it shows the **expected future damage** if nothing is done to control the causes of the damage. Please take the time to read this carefully.

STAGE A - 1970

The richness and abundance of fauna and flora indicate an ecosystem without significant environmental perturbations.

STAGE B- TODAY

Environmental changes arising from human activities have provoked changes in the Pantanal's water quality, undermining the equilibrium of the ecosystem and reducing the populations of many species of animals and plants.

STAGE C - 2010

The negative impacts of environmental degradation will lead to a rapid and continuous reduction in the populations of the most sensitive species, threatening them with extinction.

FIGURE 6.3 Scenario Card for Environmental Change in the Pantanal (See Questionnaire)

LEAVE THE CARD IN FRONT OF THE RESPONDENT, AND ALLOW TIME FOR IT TO BE EXAMINED. WHEN THE RESPONDENT HAS FINISHED READING, EXPLAIN THE FOLLOWING:

Currently, the damages are in **Stage B.** Without controls, it might be possible that **Stage C** will be reached in the year 2010. Using **new technologies** to reduce this damage, scientists hope that the environment will not degrade to **Stage C** levels.

SHOW THE PANTANAL MAP, AND LEAVE IT IN FRONT OF THE RESPONDENT.

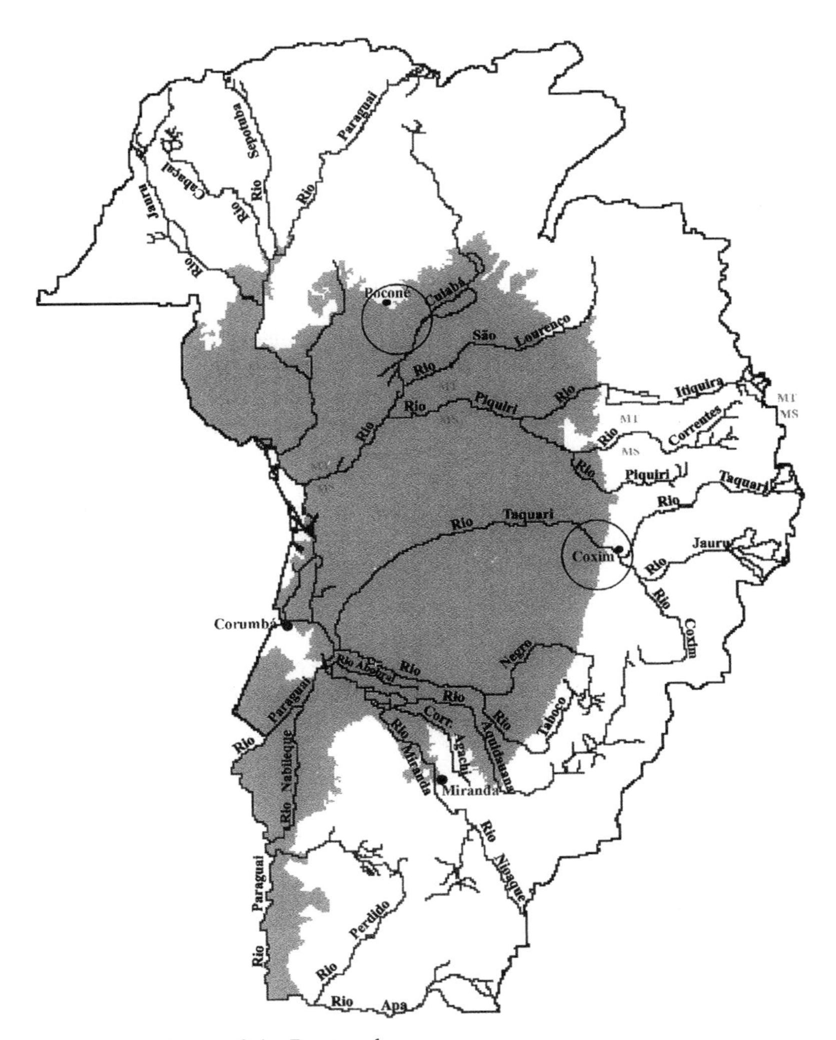

FIGURE 6.4 Map of the Pantanal

In the highlighted areas on this map, the extent of damage is **locally severe.** Severe damage may occur **in many other rivers of the Pantanal too.**

20A. Did you know about the pollution problems caused by mercury or silt in the Pantanal?

a. ___ Yes

b. ___ No

20B. If you answered YES, what was the source of your knowledge?

______

NEXT, READ ALOUD THE FOLLOWING:

Implementing controls means high costs, because the technologies used for the recovery of the environment are expensive. Some level of water quality is currently maintained in Brazil by tax revenues. Users like you also pay directly for the buildings, services, and facilities (public or private) that exist in recreation and leisure places.

However, in some places, such as the Pantanal, the revenues coming from these payments are insufficient to guarantee the maintenance of water quality. So, more money is necessary to maintain the present quality and avoid continued degradation.

One important way to induce the government to spend more money on maintenance of water quality is to find out how much the visitors value an environment with the characteristics of the Pantanal and the leisure activities that are possible there.

To get some idea of how much this is worth, we are asking people a few questions about the amount of money they would be willing to pay to ensure that the present quality of the water of the Pantanal rivers is maintained.

As you answer these questions, please keep in mind the following points:

1. Rest assured that all proceeds coming from these payments will be used **exclusively to maintain the water quality in the Pantanal.**
2. With these proceeds, it will be possible to maintain the water quality conditions (and consequently the leisure opportunities) at their present level (**Stage B**).
3. If sufficient funds cannot be provided through these payments, the water quality certainly will reach **Stage C.**
4. Base your answers on the kinds of leisure activities that are available to you today, and on those that you might enjoy in the future.
5. Remember that the amount you decide you can pay to ensure water quality will come from your household budget, so you cannot spend it on other activities.
6. All river users (including boat operators) will pay the same amount, although the form of payment may vary. These payments are valid only for the leisure activities in Mato do Grosso do Sul State.

Do you completely understand these points? Would you like me to repeat any?

DISCRETE-CHOICE QUESTIONNAIRE

21. Keeping in mind these considerations, would you be willing to pay some amount of money to maintain the water quality of the Pantanal rivers?
    a. ___ Yes (Continue with the text below)
    b. ___ No (Go to Question 26)
    c. ___ Don't know (Continue with the text below)

Let us consider the possibility of a single increase in the price of a fishing licence, currently costing R$34.00, with the EXCLUSIVE objective of obtaining revenue to invest in a system of water quality control. The following questions relate to the maximum amount you would be willing to pay for the implementation of such a system.

22. What is the **maximum** increase you might pay for the licence? You may name any amount you think is appropriate. R$ ______________

If donation was R$ = 0, go to Question 26.

23. If this amount was not enough to ensure the maintenance of water quality in the Pantanal, would you pay more (any amount more)?
    a. ___ Yes (Go to Question 24)
    b. ___ No (Go to Question 25)
    c. ___ Don't know (Go to Question 25)
24. What is the maximum **additional** amount (i.e., on **top** of the sum given in your previous answer to Question 22) that you would be willing to pay? R$ _____
25. What did you mean when you answered you would pay R$ _____ more (the amount given in question 22)? Have you determined that you are truly willing to pay this amount?_____ (Go to Question 27)
26. You are not willing or able to donate to maintain water quality. Why?
    a. ___ I cannot afford to pay, although I would like to.
    b. ___ My value = zero. Why? ______________
    c. ___ Water pollution is not important to me.
    d. ___ I refuse to value. Why? ______________
    e. ___ I feel that this is someone else's responsibility (for example, the government's)
    f. ___ I pay too much tax, etc., already.
    g. ___ No answer.

Finally, I would like to find out some characteristics of your household. This is to ensure that we have interviewed a cross section of the population for our survey.

27. This card (indicate showcard 4) lists household incomes in R$. Remembering that **all answers are *completely anonymous*,** please indicate into which of these broad groups your **total monthly household income** falls. Please include any state benefits or pensions, interest on investments, etc.; in other words, include **all** pre-tax incomes to the household. R$_
    a. ___ Less than 1,000.00
    b. ___ From 1,001.00 to 1,500.00
    c. ___ From 1,501.00 to 2,000.00
    d. ___ From 2,001.00 to 2,500.00
    e. ___ From 2,501.00 to 3,000.00
    f. ___ From 3,001.00 to 3,500.00
    g. ___ From 3,501.00 to 4,000.00
    h. ___ From 4,001.00 to 4,500.00
    i. ___ From 4,501.00 to 5,000.00

j. ___ From 5,001.00 to 5,500.00
l. ___ From 5,501.00 to 6,000.00
m. ___ From 6,001.00 to 6,500.00
n. ___ From 6,501.00 to 7,000.00
o. ___ From 7,001.00 to 7,500.00
p. ___ From 7,501.00 to 8,000.00
q. ___ From 8,001.00 to 8,500.00
r. ___ From 8,501.00 to 9,000.00
s. ___ From 9,001.00 to 9,500.00
t. ___ From 9,501.00 to 10,000.00
u. ___ Over 10,000.00

28. Could you tell me your birth year? _______________

29A. This card (indicate showcard 5) lists levels of formal education. Please choose the highest level you have completed.
a. ___ Never attended school
b. ___ Some primary education
c. ___ Primary school degree
d. ___ Secondary school degree
e. ___ Technical degree
f. ___ Baccalaureate degree
g. ___ Postbaccalaureate degree

29B. Your **profession** is? _______________

30. Finally, are you a member of any nature conservation organization?
a. ___ Yes (Please name them. ____________________________)
b. ___ No

THANK YOU FOR YOUR HELP AND ATTENTION!

Time interview completed (24-hour clock): _____ : _____ hour

DOUBLE-BOUNDED FORMAT QUESTIONNAIRE

21. Keeping in mind these considerations, would you be willing to pay some amount of money to maintain the water quality of the Pantanal rivers?
a. ___ Yes (Continue with the text below)
b. ___ No (Go to Question 26)
c. ___ Don't know (Continue with the text below)

By **conservative** estimate, the cost of the technology needed to maintain the water quality level in the Pantanal at **Stage B** will require an increase in the price of a fishing licence of R$ _ [to be obtained prior to the interview].

22. Would you be willing to pay this **additional** amount for the fishing licence (**above** the current price of R$34.00)?
a. ___ Yes (Go to Question 23A)
b. ___ No (Go to Question 23B)
c. ___ Don't know (Go to Question 23B)

23A. We still don't know exactly how much of an investment will be necessary to maintain the water quality level in the Pantanal. If the **final** estimated cost of the cleaning technology is **R$** _____ (double the amount suggested previously), would you be willing to pay this **additional** amount for the fishing licence (**above** the current price of R$34.00)?

a. ___ Yes (Go to Question 24)
b. ___ No (Go to Question 24)
c. ___ Don't know (Go to Question 24)

23B. We still don't know exactly how much of an investment will be necessary to maintain the water quality level in the Pantanal. If the **final** estimated cost of the cleaning technology is **R$** _____ (half the amount suggested previously), would you be willing to pay **this additional** amount for the fishing licence (**above** the current price of R$34.00)?

a. ___ Yes (Go to Question 24)
b. ___ No (Go to Question 24)
c. ___ Don't know (Go to Question 24)

24. What is the maximum **additional** amount (i.e., **above** of the actual price of R$34.00) that you would be willing to pay for the fishing licence? R$ _______________

THANK YOU FOR YOUR HELP AND ATTENTION!

Time interview completed (24-hour clock): _____ : _____ hour

NOTES

1. Opposing views on the debate can be found in papers by Hanemann (1994), Portney (1994), and Diamond and Hausman (1994), in a special issue of the *Journal of Economic Perspectives.*
2. The ecosystem is complex in the sense that it contains use and nonuse components and goods remote from a consumer's everyday experience.
3. The survey used in this study is attached in the appendix.
4. It is important to note that the panel guidelines are still the subject of considerable disagreement. Unconditional adoption of every aspect in design is in most cases inappropriate.
5. Logit and probit models—which in fact should return similar parameter values—are typically used because they are conveniently offered in most econometric packages such as LIMDEP.
6. Values $-1 < \beta < 0$ in fact lead to an infinite mean, while values of β less than -1 still lend considerable weight to the right tail of the distribution.
7. An alternative strategy is to use the median WTP, which is less sensitive to the tails of the distribution. However, the median implies a different social welfare criterion based on the median voter.
8. The exact wording of scenarios and WTP questions (open-ended and discrete-choice) may be found in the appendix.
9. Over the period of the survey (July to November 1994), the Brazilian real stood at approximately R$1 = US$0.9.

10. "What is the maximum amount of money you/your household would be willing to pay per year to have/avoid this change?"
11. Catches were weighed for the issue of lacre.
12. This figure relates only to the state of Mato Grosso do Sul and is based solely on the issue of lacre by the Forest Police. It probably underestimates the number of actual anglers. At least an equal number of visitors might be assumed for Mato Grosso state.
13. We thank Joe Cooper of the USDA Economic Research Service for providing a copy of the DWEABS algorithm.
14. For a review of optimal design methods used in biostatistics, see Alberini (1995) and Cooper (1993).
15. Where the effect of omitted variables shows up only as increased residual deviance, provided omitted variables are uncorrelated with retained regressors.
16. *Know* indicates whether the respondent admitted to having prior knowledge of the pollution problem.
17. For the unrestricted (multivariate) model, the correct term for the unrestricted mean is $1/\beta(\alpha + \Sigma\gamma_i s_i)$, where γ_i is the estimated parameter corresponding to the socioeconomic characteristic s_i, which is evaluated at its mean.
18. Strictly speaking, the second response is conditional on the first, so it is unlikely that response motives will be entirely unrelated.
19. In the range from −13.04 to 1143.71.
20. Derived by treating nonrespondents and item nonresponses as no responses to set amounts.
21. All but the nonparametric estimate are also bounded by the open-ended bid range.

7

Estimation of Water Quality Control Benefits and Instruments in Brazil

This chapter treats questions associated with urban-industrial water pollution and the costs to human settlements exposed to risk as a consequence. Environmental economist Serôa da Motta and collaborators offer two case studies of valuation appropriate to the determination of investment requirements and internalization of pollution damages.

The assessment of health damages arising from exposure to pollution represents a widely used approach to the valuation of environmental damages. The assessment of waterborne disease damages estimated by Serôa da Motta and Rezende in the first case below derives statistical relationships associating mortality with the lack of basic sewerage and adequately treated water supplies, common in many parts of Brazil. Monetary values are then derived based on the imputed number of lives saved for a given marginal investment in sanitation services, a value that can be of great significance for investment choices.

In the second case in this chapter, Serôa da Motta and Mendes assess the marginal costs attributable to water pollution control investments by industrial facilities in the Tietê River basin in São Paulo, as a means to simulate the effectiveness of effluent taxation on technology adoption. They assume here that firms able to reduce emissions at a cost-per-unit-emitted lower than the tax rate would do so, while those with more costly pollution control would be better off paying the tax. The simulation in this case study shows that the social costs of such market-based measures would be considerably lower than those associated with a uniform emission ceiling across firms.

—P. H. M.

The Impact of Sanitation on Waterborne Diseases in Brazil

RONALDO SERÔA DA MOTTA
and LEONARDO REZENDE

The expansion of the supply of public treated water and the provision of systems of sewage collection and treatment to the urban population in developing countries have been often pointed to as one of the major steps toward higher living standards and environmental quality.

It is widely recognized that one important aspect of the lack of sanitation services is the incidence of waterborne diseases that are responsible for a high rate of child mortality.

This case study aims to analyze the role of sanitation services in reducing the mortality rate from waterborne diseases in Brazil. It develops a model to test the statistical relationship between these mortality rates and the provision of sanitation facilities. Based on these results, one will be able to identify the importance of each type of sanitation service in mortality reduction and, consequently, determine priority of policy actions. As will be seen, although the impact of water supply is still very significant, the expansion of sewage systems, due to their current absence, will also make a major contribution to the control of mortality due to waterborne diseases in Brazil. The last section will deal with the investment costs required to achieve a higher level of sanitation service and the resulting cost per life saved. These values represent reasonable indicators of the consumption that Brazilian society must forego to offer adequate sanitation services and, mainly, the efficiency at which the sanitation sector is currently operating. It will then become clear whether water supply costs for each life saved justify this service as a priority in sanitation policy, and furthermore whether jointly undertaking sewage services with water supply seems to be an economically sound option. Nevertheless, sanitation investment costs in Brazil are very high, which seems to cast some doubt on the efficiency of the current provision of sanitation services in the country.

WATERBORNE DISEASES AND SANITATION IN BRAZIL

In this study, the six most common waterborne diseases were selected to be analyzed: intestinal infections, cholera, typhoid fever, poliomy-

TABLE 7.1 MORTALITY CASES ASSOCIATED WITH WATERBORNE DISEASES IN BRAZIL — 1981 TO 1989

	Mortality Cases (% of total cases)			
	Intestinal Infections		*Other*[a]	
Age	1981	1989	1981	1989
<1 yr	28,606 (81.8)	13,598 (72.0)	87 (9.4)	19 (2.9)
1–14 yr	3,908 (11.2)	1,963 (10.4)	44 (4.8)	21 (3.2)
>14 yr	2,439 (7.0)	3,330 (17.6)	793 (85.8)	608 (93.8)

From Ministry of Health, Brazil.
[a]Cholera, typhoid fever, poliomyelitis, amoebic dysentery, schistosomiasis, and shigellosis.

elitis, amoebic dysentery, schistosomiasis, and shigellosis, for which the most recent mortality figures are presented in table 7.1.[1]

As can be seen, the number of cases has dropped sharply in the period from 1981 to 1989. Nevertheless, intestinal infections in children still represent 80 percent of the cases in 1989. It is worth mentioning that 1989 data do not capture the very recent cholera epidemic period, which resulted in almost 100 deaths in 1992 alone.

The relevant literature on health issues in developing countries[2] clearly indicates that sanitation conditions are one of the most important causes of waterborne diseases. That situation is due to the large investments needed to provide adequate sanitation services vis à vis the low investment capacity of these countries. For instance, in the case of Brazil (Serôa da Motta and Mendes 1993), the investment needed to offer adequate sanitation services to the unserved population would be between 11 and 19 billion dollars, varying according to the level of sewage service. Moreover, the adopted sanitation policies have been designed and implemented without the definition of clear and sound economic criteria to define goals and evaluate service performance.

Although the provision of sanitation services in urban areas in Brazil is still inadequate, it has significantly increased during the eighties,[3] particularly if one considers the public investment restrictions in that period. Using the most recent information, it can be seen in table 7.2 that water supply already covers 89.4 percent of the urban population, up from 78.4 percent in 1981. Another large change is observed for sewage collection, whose 1981 coverage was

TABLE 7.2 PROVISION OF SANITATION SERVICES IN BRAZIL — 1981 TO 1989

Households per Income Group	*Proportion of the Population with Sanitation Services by Income Group (%)*					
	Treated Water		*Sewage Collection*		*Sewage Treatment*	
	1981	*1989*	*1981*	*1989*	*1981*	*1989*
0–2 MW	59.3	76.0	15.0	24.2	0.6	4.7
2–5 MW	76.3	87.8	29.7	39.7	1.3	8.2
>5 MW	90.7	95.2	54.8	61.2	2.5	13.1
All	78.4	89.4	36.7	47.8	1.6	10.1

From IBGE: Household surveys (PNAD) and sanitation surveys (PNSB).
MW, multiples of the minimum wage (which varied during the period between US$50 and US$100 per month).

only 36.7 percent and in 1989 reached 47.8 percent. Sewage treatment coverage also expanded from less than 2 percent in 1981 to over 10 percent in 1989.

However, when the income distribution of these services is analyzed, it is easily recognized that the poor are those who have by far the lowest coverage. In 1989, for public water and sewage, respectively, the figures in table 7.2 show that service was extended to less than 70 percent and 20 percent of the lowest income bracket [< 2 times the minimum wage (MW)]. Coverages increase to 76 percent and 48 percent for households that earned between 2 and 5 times the minimum wage. For those earning more than 5 times the minimum wage, provision is much higher, at 91 percent for water and 76 percent for sewage systems.

Although data on mortality are not available according to income level, poor children can be expected to be the principal victims of waterborne diseases, since their inadequate sanitation conditions increase exposure. They cannot combat this with defensive expenditures because of their households' income restrictions.

THE MODEL

An important issue in environmental economics is the estimation of dose-response functions to evaluate the underlying welfare changes arising from environmental damage. Dose-response estimates related to air pollution are fairly common in the literature [see, for example,

Ostro (1992) and Serôa da Motta and Mendes (1993)]. Estimation of the impact of water pollution on health, however, has attracted considerably less interest from researchers, probably because deficient sanitation has already been eliminated in most developed countries. Unfortunately, this is not the case for Brazil. In this case study, an attempt is made to estimate a function relating sanitation conditions and mortality due to waterborne disease using Brazilian data for the 1980s.[4]

Here we apply a PROBIT model, since the observed mortality rates are considered a sum of dummy trials. Techniques dealing with dummy dependent variables are common in environmental economics literature, mostly in contingent valuation models. It will be asserted here that this technique can also be useful in dose-response models.

Model Presentation

It is here assumed that water pollution affects human health by increasing the probability of people becoming ill or dying due to waterborne diseases. The mortality rate is not, however, a direct measure of health. Death is assumed to occur when an unseen variable exceeds a threshold value. This variable is a true measure of the likelihood of being affected by waterborne diseases and it is indirectly observed by the registered mortality rates. By assumption, the unseen variable is a linear function of the independent variables.

Therefore, this research intends to test an econometric model logically divided into two steps. First, the model assumes the probability distribution of an individual's dying, which will be represented by a critical value (z). Thus z is the argument of a standardized normal function that estimates the probability of an individual's dying. More detailed description of the model is presented in the appendix to this chapter.

Two relevant hypotheses are here assumed:

1. A normal distribution is adopted for the probability functions. The best reason for this is the central limit theorem; if one supposes that this event is characterized by a sum (or an average) of many random variables, this theorem states that, regardless of their particular individual distributions, the sum will follow a normal one. Another very popular distribution that could be used is the logistic, which approximates quite well the normal distribution and is algebraically much simpler. It nevertheless has thicker tails, which can be a major drawback as probabilities near zero will be modeled. Moreover,

as the computing resources allow estimation with the normality assumption, simplicity is no reason to use the logistic distribution.[5]

2. The events modeled are considered independent from the outcomes of other trials. This means that the probability of an individual's dying is not altered by what happens to his neighbors. No epidemic behavior is considered, which is a very plausible assumption with respect to the diseases being analyzed.

Description of Variables and Database

Theoretically, the model considers each individual at each year as an independent observation. However, then y would be a vector of zeros and ones with about 1.2 billion lines! Furthermore, the dependent data set cannot be divided into that much detail. Bearing this in mind, these observations can be grouped by state and by year. Using these groups instead of the individual trials reduces the number of variables to about 205 with clear econometric benefits. A consistent estimate of the true probability of dying can now be computed as P = number of deaths per total population. The dependent variable to be put in the regression is the mortality rate observed for each state.

Then X is a database for each Brazilian state for the period from 1981 to 1989. The variables considered in the model are the following:

PRG, percentage of the urban population supplied by public water system
PCEFS, percentage of the urban population connected to the public sewer system or using septic tanks
PFIL, percentage of the urban population using domestic water filter
PLIXO, percentage of the urban population with solid waste collection services
PLEITOS, number of hospital beds per capita.

Dummy variables for each state were added. This procedure is equivalent to decomposing each residual u_{ij} into a sum $E_i + u_{ij}$, with i being a state index and j standing for each year. E_i refers to the impact of many geographical, social, climate, hydrological aspects not directly considered in the model. This is an advisable procedure[6] when estimating over pooled data, since more noise is supposed to be found between different states than along different years.

Sanitation and waste collection data were obtained from a special tabulation of household and sanitation surveys and hospital beds from health sector inquiries. In the case of sanitation and waste collection, the database allows discrimination by income level. Therefore, three X sets were tested: the whole urban population, urban population earning less than 5 times the minimum wage, and urban population earning less than 2 times the minimum wage.

Mortality rates are obtained from death certificates, computed by the Health Ministry, covering the following diseases: cholera, typhoid fever, food poisoning, shigellosis, amoebiasis, poliomyelitis, schistosomiasis, and intestinal infections. This database has the advantage of not being the result of sampling projection, but information is not available by income levels. However, as has been discussed, these diseases affect principally the poorest people.

It was possible to aggregate the number of deaths by the age of the victim. Three groups were considered: all ages, children between birth and fourteen years old, and children younger than one year old. Both dependent and independent variables considered were available for all twenty-five Brazilian states yearly from 1981 to 1989. The pooled database has therefore 205 observations.

Estimation Method and Results

Table 7.3 presents the coefficient estimates according to income levels and distinct set of variables. Values in brackets are the t-tests. Except for PLEITOS, all coefficient signals conform to the theory: the higher the level of sanitation, the lower are the cases of mortality.

The variable PLEITOS was introduced to represent a measure of medical assistance available, which could then explain part of the spatial variation in death rates and correct a bias caused by the fact that greater disease exposure can be overcome by efficient medical care. Therefore, a larger number of beds per capita was assumed to result in a reduced death probability. However, estimated PLEITOS coefficients are, in all models, significantly positive. One can consider then that the positive coefficient tells us that more death cases requires a stronger medical effort and, therefore, that variable may not be indicated in this model. When PLIXO is not considered, the PRG coefficient increases and becomes more significant. This fact is worth noting, since PRG is theoretically supposed to have a more direct relationship with waterborne diseases.

Considering only children, one should expect to find greater coefficients since the dependent variable (mortality cases) largely increases. This indeed happened, as shown in table 7.4, with the exception of PFIL and PLIXO. More specific trends can be observed when taking only sanitation conditions of the poor. For example,

Table 7.3 Coefficient Estimation of Overall Mortality, by Income Level

	Variable						
Income Level	*PRG*	*PCEFS*	*PFIL*	*PLIXO*	*PLEITOS*	*R-square*	*Adj. R-square*
Total							
	−0.111	−0.175	−0.422	−0.417	70.149	91.5	90.3
	(−0.825)[a]	(−1.718)	(−2.534)	(−2.857)	(−7.196)	—	—
	−0.122	−0.306	−0.371	−0.707	—	90.7	89.4
	(−0.808)	(−2.766)	(−1.986)	(−4.481)	—	—	—
	−0.530	−0.438	−0.570	—	—	88.4	86.7
	(−4.214)	(−3.943)	(−2.986)	—	—	—	—
<5 MW							
	−0.189	−0.218	0.041	−0.359	41.787	87.6	85.7
	(−1.621)	(−2.549)	(−0.265)	(−2.955)	(−4.022)	—	—
	−0.249	−0.295	0.099	−0.436	—	86.5	84.6
	(−2.068)	(−3.409)	(−0.613)	(−3.496)	—	—	—
<2 MW							
	−0.216	−0.318	0.018	−0.243	65.469	90.7	89.2
	(−2.232)	(−3.833)	(−0.150)	(−2.219)	(−6.347)	—	—
	−0.319	−0.473	−0.061	−0.327	—	88.7	87.0
	(−3.023)	(−5.405)	(−0.452)	(−2.732)	—	—	—

PRG, percentage of the urban population supplied by public water system; PCEFS, percentage of the urban population connected to the public sewer system or using septic tanks; PFIL, percentage of the urban population using domestic water filter; PLIXO, percentage of the urban population with solid waste collection services; PLEITOS, number of hospital beds per capita; MW, multiples of the minimum wage.

[a]Statistic *t* in parentheses.

Table 7.4 Coefficient Estimation of Child Mortality, by Income and Age Groups

Age of Victims	Income Level	Variable					R-square	Adj. R-square
		PRG	PCEFS	PFIL	PLIXO	PLEITOS		
0–14 yr	Total	−0.192	−0.218	−0.567	−0.528	79.337	89.5	88
		(−1.249)[a]	(−1.828)	(−2.888)	(−3.142)	(6.878)		
	<5 MW	−0.281	−0.296	0.157	−0.479	86.983	89.0	87
		(−2.255)	(−3.094)	(0.898)	(3.650)	(7.674)		
	<2 MW	−0.303	−0.381	0.007	−0.331	75.036	89.0	87.6
		(−4.294)	(−1.052)	(0.346)	(−0.355)	(3.069)		
0–1 yr	Total	−0.250	−0.279	−0.413	−0.377	83.064	87.5	85.7
		(−1.226)	(−1.778)	(−1.602)	(−1.712)	(5.664)		
	<5 MW	−0.337	−0.332	0.356	−0.326	88.448	87.5	85.6
		(−2.096)	(−2.720)	(1.573)	(1.929)	(6.240)		
	<2 MW	−0.359	−0.438	0.083	−0.170	74.383	87.8	85.9
		(−2.444)	(−3.505)	(0.446)	(−1.033)	(4.834)		

PRG, PCEFS, PFIL, PLIXO, PLEITOS, and MW, see table 7.3 notes.
[a]Statistic t in parentheses.

significance is reduced in PFIL coefficients suggesting that the presence of filtering facilities alone among the rich has an impact on their mortality rates. PRG and PCEFS, as theoretically expected, increase, and PRG in particular gains significance. Inversely, the PLIXO coefficient decreases.

All these conclusions are confirmed if maximum likelihood estimation is used when, again, most of the unexpected behavior came from PFIL and PLIXO—the variables supposed to have a less important effect on the diseases' incidence. This suggests that the significance of these variables cannot be taken for granted since adult death and sanitation conditions among the rich bring noise to the model.

Further simulations were carried out to confirm some hypotheses drawn from the regression results in two ways:

1. by splitting up the regression into time series and cross sections, and
2. by introducing distinct measurements of variables.

First, an analysis was conducted of the difference between these results and those presented by Serôa da Motta and Mendes (1993), where a preliminary dose-response function for waterborne diseases was determined, and incidence was significantly affected only by sewage systems. Here, however, water as well as sewage systems were found to be significant at approximately the same level. This is explained by the introduction of the time dimension in the cross-section database: in Serôa da Motta and Mendes (1993), data for only 1988 and 1989 were utilized. When, using the nine-year database, regressions are estimated for PRG and PCEFS separately by state and year, and coefficients are obtained for twenty-five distinct time series and nine cross sections. For cross sections, PCEFS was found to be always negative and statistically very significant, while PRG was never significantly far from zero. Inversely, for the time series, PRG was considered significantly negative in ten states including the larger ones, whereas that occurs to PCEFS only in small states. This test suggests therefore that the explanatory power of PRG is concentrated on change within each state as time passes, while PCEFS explains differences among states. This is why Serôa da Motta and Mendes (1993), dealing with a database for only two years, found that only sewage systems were significant.

The contradictory behavior between PRG and PLIXO (see results in table 7.5) was also investigated. The overall correlation coefficient between them, for those earning less than 2 times the minimum wage,

TABLE 7.5 ALTERNATIVE REGRESSION FUNCTIONS OF PROBIT Z[a]

PRG	*PAGUA*	*PCEFS*	*PCETR*	*PETR*	*PFIL*	*PLIXO*	*PLEITOS*	*R-square*	*Adj. R-square*
−0.299	—	−0.532	0.302	—	0.213	−0.430	80.076	89.38	87.69
(−2.421)[b]	—	(−4.162)	(2.720)	—	(1.229)	(−3.302)	(7.033)		
−0.245	—	−0.229	—	−0.449	0.184	−0.485	85.195	89.45	87.78
(−1.988)	—	(−2.365)	—	(−2.523)	(1.072)	(−3.765)	(7.490)		
−0.242	—	−0.590	0.520	−0.914	0.298	−0.405	72.336	90.40	88.82
(−2.044)	—	(−4.857)	(4.522)	(−4.536)	(1.793)	(−3.269)	(6.472)		
−0.370	—	—	−0.026	—	0.128	−0.563	93.170	88.53	86.78
(−2.917)	—	—	(−0.311)	—	(0.716)	(−4.271)	(8.257)		
−0.313	—	—	—	−0.556	0.134	−0.528	90.808	89.18	87.57
(−2.598)	—	—	—	−3.237	(0.791)	(−4.227)	(8.170)		
−0.331	—	—	0.147	−0.745	0.186	−0.552	88.559	89.30	87.60
(−2.691)	—	—	(1.624)	(−3.598)	(1.073)	(−4.345)	(7.891)		
−0.287	—	−0.2904	—	—	—	−0.4536	87.814	89.06	87.45
(−2.305)	—	(−3.058)	—	—	—	(−3.534)	(7.790)		
−0.05373	−0.578	−0.298	—	—	—	−0.3529	88.8601	89.39	87.77
(−0.292)	(−1.851)	(−3.177)	—	—	—	(−2.613)	(7.967)		
	−0.6448	−0.3028	—	—	—	−0.3579	89.26	89.38	87.82
	(−3.075)	(−3.289)	—	—	—	(−2.676)	(8.082)		
−0.729437	—	−0.46761	—	−0.6219	—	—	—	85.25	83.18
(−6.333)	—	(−4.327)	—	(−2.957)	—	—	—		

PRG, PCEFS, PFIL, PLIXO, PLEITOS, and MW, see table 7.3 notes. PAGUA, the part of the population drinking adequate water; PCETR, the part of the population whose sewage has been collected and treated by either the public system or septic tanks; PETR, the part of the population whose sewage has been collected and treated by the public system.

[a]For earnings lower than 5 times the minimum wage and children younger than 14 years.

[b]Statistic t in parentheses.

was estimated to be 0.64. However, when the database is segregated into time series and cross section, as explained earlier, the correlation increases to around 80 to 90 percent. This means that PLIXO and PRG move together. The same behavior was found in other samples of income level groups. Since PRG has a strong explanatory power, the variable PLIXO should be left out to preserve PRG significance.

Another finding is related to the positive PLEITOS coefficient. Once more, the technique of segregating the database into time series and cross sections proved to offer meaningful evidence. Fitting $z = f(\text{PLEITOS})$ by states, significantly positive coefficients were found. However, when one considers cross sections, the coefficient becomes significantly negative in all years, as theoretically expected. The unexpected fitting over time can be explained by specialists. During the decade, while death rates decreased, hospital facilities, along with many other social indicators, deteriorated as a result of the long public sector crisis. Therefore, introducing PLEITOS is only harmful and it would be useful only in the cross section dimensions—but in that case the dummies can capture its influence.

Another variable was unsuccessfully tested to measure hospital facilities in place of PLEITOS. A new variable, PPOCUP, was defined as the per capita number of people trained at the undergraduate level and employed in hospitals. When estimated together with PLEITOS, PPOCUP is found not to be significant, and when estimated without it, it occupies the role of PLEITOS, yielding a strongly significant (and unexplainable) positive coefficient.

A new variable PCETR was also introduced as the percentage of the population whose sewage has been collected and treated by either public system or septic tanks. In addition, another variable was considered, PETR, where only public systems are counted. As can be noted in table 7.5, PETR showed, apart from correct signals, more significant coefficients than PCETR.

Finally, there was also a problem about the framework used to estimate water supply influence. The relevant issue here is the quality of water consumed by the population which is influenced by both PRG and PFIL. Therefore, PAGUA was introduced as another variable representing the part of the population drinking adequate water (i.e., filtered water from any source or water coming from the public system, whether filtered or not). The resulting coefficients (see table 7.5) did not prove better than PRG, which may reflect the higher importance of treated public water.

TABLE 7.6 EFFECT OF INCREASING ACCESS TO SANITATION SERVICES BY 1%, TO THE POPULATION OF BRAZIL IN 1989

	Sanitation Services[a]			
	Treated Water	*Sewage Collection*	*Sewage Treatment*	*All Three Services*
Children's Lives Saved[b] (% of total mortality cases)	463 (2.5)	298 (1.6)	395 (2.1)	1,133 (6.1)
Cost of Each Life Saved[c] (US$)	115,102	214,562	175,207	164,385

[a]In the first three columns, the named service was expanded without changing the others; in the last column, all three were expanded together.
[b]Deaths counted were those associated with waterborne diseases (see table 7.2).
[c]Total cost to increase access by 1%, divided by number of lives saved.

THE ROLE OF SANITATION IN WATERBORNE DISEASES

The lack of sanitation services affects mostly poor people who cannot afford defensive expenditure. Moreover, as discussed earlier, children have the highest probability of dying from exposure to waterborne diseases. Therefore, to assess in detail the importance of sanitation in reducing mortality rates due to waterborne diseases, the regression for income levels of less than 5 times the minimum wage and children under 14 years of age was selected.

The variables considered are those that were revealed to be more significant as discussed in the last section: PRG (public treated water supply), PCEFS (sewage collection system), and PETR (sewage treatment system).

Using the regression, it is possible to estimate the number of children who would probably be saved if the supply of sanitation services is expanded by 1 percent of the population of the families earning less than 5 times the minimum wage, who were not provided with sanitation services in the year 1989.

As can be seen in table 7.6, increasing the provision of public water supply by 1 percent of that population not yet served, without any improvement in other sanitation services, will probably reduce by 2.5 percent the number of mortality cases of children under 14 years of age associated with waterborne diseases (462 lives). For a 1 percent increase in sewage collection and treatment, the equivalent reductions would be, respectively, 1.6 percent (298 lives) and 2.1 percent (395 lives). When all three services are jointly provided for that population, the reduction levels would be about 6.1 percent (1139 lives). The importance of water supply is confirmed by its higher reduction effect resulting from an initial 1 percent increase in service expansion.

It should be noted that the contribution of all services together is not exactly the sum of all the services' contributions, since a normal distribution is considered. Therefore, minor differences can be found without revealing changes in the explanatory capacity of the variables.

However, it is worth noting that complementarity is expected among sanitation services regarding water quality improvement and the resulting decrease in cases of waterborne diseases. That possibility results from the fact that sewage collection diverts effluent discharges from streets and channels and, consequently, avoids contact with polluted water. In turn, sewage treatment reduces the risk of contamination of water supply from reservoirs as well as from point sources serving those not connected to public systems. Therefore, one must analyze the economic efficiency of enhancing this complementarity when sanitation investments are undertaken.

The investment required for each life saved from a 1 percent increase in the served population is also presented in table 7. 6. These costs are determined as the total investment necessary to increase by 1 percent the served population for each sanitation service, divided by the total lives saved by that increase. Costs are estimated according to average per capita investment costs of the conventional sanitation service projects undertaken by the sanitation companies in Brazil.[7] Lives saved are those calculated earlier.

As can be seen in table 7.6, the cost for each life saved related to public water supply is $115,102, followed by sewage treatment and collection costs that are, respectively, $175,207 and $214,562. When all three services are jointly expanded by 1 percent, the Brazilian society should invest an amount of $164,385. The lower costs of water services seem to present an economic justification of the option, historically taken in Brazil, to give priority to water services at the expense of sewage systems. However, the expansion of all three services together increases the probability of saving lives, which, although still more costly than the alternative of water supply alone, will represent a lower investment value for each life saved than the case of sewer system expansion in a distinct time path. This possibility can be explained by the fact that water and sewage treatment costs present higher mortality probabilities and lower costs than sewage collection and, consequently, lead to a lower average value. Moreover, the prospect that water supply may be expanded is limited, since that service already covers almost 84 percent of the poor population.

Summing up, the effect on mortality reduction in children of low-income families obtained by the provision of sanitation facilities is very significant in the cases of waterborne diseases. The expansion of publicly treated water systems is a key factor to reduce mortality whereas, because of their current shorter supply, sewage collection and, mainly, treatment services will become more important to secure higher levels of sanitation coverage. The sanitation investments needed to reduce these deaths indicate that water supply is a priority because of its lower cost for each life saved, although these costs for sewage systems are largely reduced when undertaken jointly with water supply investments.

These costs may be regarded as the value the Brazilian society ascribes to the possibility of saving one poor child's life. Considering that sanitation facilities last for more than fifty years and operational and maintenance costs are equivalent to 10 percent of capital costs, the annualized cost of the saved life would be, for the joint alternative, around $18,000, which is more than 4 times the Brazilian per capita income. The actual investment by Brazil in joint basic sanitation remains low when one considers the importance of saving lives and the relatively low investment requirements. These figures indicate that the impacts of sanitation services on waterborne diseases can be measured and evaluated using economic criteria, offering an indicator of service deficiency and the benefits of public investment.

Environmental Standards, Revenue Generation, and Pollution Taxes: A Simulation for Tietê River Basin in Brazil

RONALDO SERÔA DA MOTTA
and FRANCISCO EDUARDO MENDES

The use of market-based instruments (MBs), such as pollution taxes, has been advocated to be a more efficient approach to environmental policy than the traditional command and control instruments

(C&Cs) applied worldwide. Moreover, MBs have also been regarded as an important instrument of revenue raising to promote funding for sustainable activities. However, the implementation of MBs is not trivial and, apart from institutional and legal aspects, issues related to their integration with the prevailing environmental standards and the resulting revenue need to be carefully examined.

This chapter develops a model to simulate the application of industrial water pollution taxes in the Tietê river basin in São Paulo state in Brazil, location of one of the most concentrated industrial sectors in the country. As will be briefly discussed in the following section, the main aim of MBs is to achieve, at the least social cost, a reduction of total pollution, say, in a basin, imposing different levels of control over the plants that contribute to overall pollutant discharge in the area. Consequently, firms decide their individual abatement level according to the difference between tax and marginal control costs.

The model presented in the second section generates results for the typical C&C when all plants are compelled to abate according to the target set by the environmental authority (i.e., environmental standards), and also the estimates for MBs when plants can choose either to abate or pay tax on emissions below the legal target.

Results discussed in the last section will indicate that targets must be reduced when the tax increases, to achieve a certain final abatement level for the basin as a whole. As the target is a proxy for standards, policies seeking fiscal revenues must set a lower tax with high target levels. Since a high tax will make pollution abatement investments viable in those heavy polluting firms having high marginal abatement costs, the size of fiscal revenue is not only linked to the level of taxation but is a result of a combination of tax and target levels. As will be analyzed, a higher target with lower tax levels alters the sectoral distribution of abatement. Consequently, the distribution of costs and revenues will also be different.

The conclusions will confirm that the decision to introduce a market-based instrument certainly leads to lower abatement costs than those observed in a command and control approach. However, if a certain level of final abatement for the basin is determined, policy makers will have to take into account that environmental standards to be imposed on individual plants are also a key parameter to define the level of abatement and the respective costs.

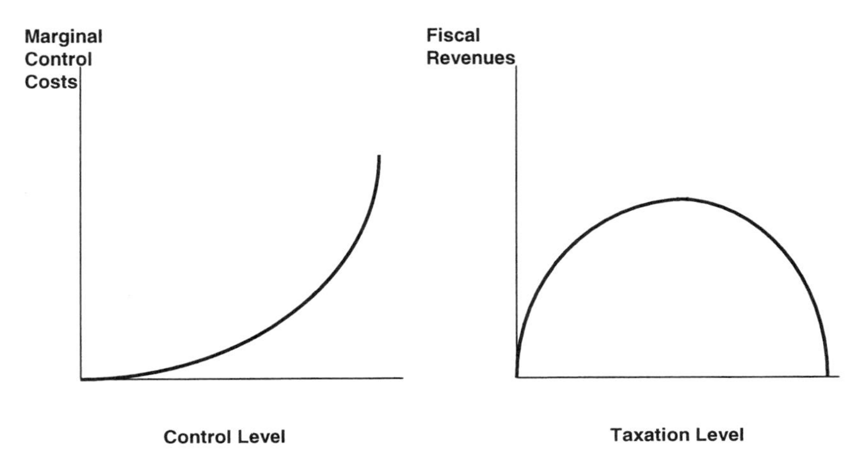

FIGURE 7.1 Abatement Cost and Fiscal Revenue Curves

MARKET-BASED INSTRUMENTS

In the command and control approach, all firms are forced to abate their potential emissions to a given percentage in order to achieve a desired overall pollution level; otherwise, they face sanctions.

A penalty, such as a tax on emission levels, introduces a component in the firm's price structure. Therefore, the decision to reduce emissions falls to the polluter, who will need to minimize the costs associated with pollution reduction. Consequently, a firm will decide to reduce its emissions until the point where the abatement cost is equal to the tax and, after that point, it will be more rational to pay the taxes. With MBs, society thus reaches abatement targets in a more efficient way at lower control costs. Total emission control is equivalent to that achieved in the C&C approach, but its distribution among polluters will differ.[8]

Market-based-instrument private costs associated with pollution—those accruing to polluters—are then basically a combination of two factors: (1) the pollution abatement costs and (2) the resources needed to pay taxes for the residual pollution that goes unabated, which can serve as a fiscal revenue for the regulatory agency. For society as a whole, MB social costs are only the control costs, since fiscal revenues represent tax payments and can be considered transfers among economic agents.

As presented in figure 7.1, the abatement cost curve has an upward slope, indicating that marginal costs increase when abatement levels increase. The fiscal revenue curve has a different shape, as also presented in figure 7.1. Fiscal revenue is expected to increase when the tax level is not high enough to induce substantial abatement, since marginal abatement costs exceed tax payment. Above a certain tax level, abatement increases and revenue decreases to zero at the point where the tax is high enough to justify total control of emissions.

Taxes, however, may not be levied on every emission level. Society usually defines environmental standards as equivalent to those that would be imposed through C&Cs, and taxation is imposed only on emissions exceeding those standards.

A MODEL FOR SIMULATING MARKET-BASED APPLICATION

The model developed to analyze MB application presents two routines: one for a typical C&C and the other for the MB.

The simulation will deal with the pollutant *organic matter* measured as biological oxygen demand (BOD). The case study will be undertaken for the upper Tietê river basin in the state of São Paulo, Brazil, for the year 1992.

Emission data, obtained from the São Paulo state environmental agency Cetesb, consist of information on the plants' main activity (sectoral classification), potential emissions (pollution load technically expected from plants without any abatement device) and residual emissions (load remaining after the implementation of abatement devices), and factory location (sub-basin, municipality) for each major plant in the river basin.[9] Marginal abatement costs for each main activity and scale of pollutant emission come from a survey of European abatement costs presented in Jantzen (1993).

Plants can reduce their emissions by installing individual control units that remove pollutants from their effluents.[10] These units use a sequence of different technologies for each main activity, as shown in table 7.7, with specific abatement efficiency and unit costs depending on the emission scale. Simulations estimate the resulting abatement level in the basin, abatement costs, and fiscal revenue.

The Command and Control Approach

To simulate a C&C control strategy, a very simple approach was adopted:

1. On all factories, from all activities, an abatement target is imposed by the environmental agency. If the abatement target is, for

Table 7.7 Technologies Adopted by Plants in the Model for Pollution Removal

Main Activity	*Technology 1*	*Technology 2*	*Technology 3*	*Technology 4*
Pulp, Paper	Water recycling	Closed cycle	Mechanical treatment	Activated sludge
Chemical	Anaerobic treatment	Activated sludge	Activated charcoal filter	—
Pharmaceutical	Anaerobic treatment	Activated sludge	Activated charcoal filter	—
Textiles	Water recycling	Mechanical treatment	Activated sludge	Activated charcoal filter
Food Products	Anaerobic treatment, high-concentration	Anaerobic treatment, low-concentration	Activated sludge	—
Beverages	Anaerobic treatment, high-concentration	Anaerobic treatment, low-concentration	Activated sludge	—
Other Industries	Mechanical treatment	Activated sludge	Activated charcoal filter	—

From Jantzen 1993.

example, 70 percent to be adopted for every plant, each plant will then abate at least 70 percent of its potential emission.[11]

2. All plants accept the regulation and invest in abatement, regardless of their size, activity, output or location.

3. Marginal abatement costs are given for each technology feasible in a given order for each main activity and are assumed equal for every plant.

The Market-Based Approach

Market-based simulation is also very simple:

1. The environmental agency sets a tax level that will be charged for each unit of pollution emitted in excess of the abatement target. This target, as for C&Cs, is a percentage of the plant's potential emission that must be either abated or taxed.
2. The same tax is imposed on every plant in all activities.
3. Each plant decides to undertake further control when marginal abatement cost is lower than marginal tax payment.
4. Marginal abatement costs are given for each technology feasible in a given order for each activity and are assumed to be equal for every plant.
5. All plants invest in abatement and/or pay the tax, regardless of their size, activity, output, or location.

This model is not a pure MB process, since there is a pollution allowance with no tax cost. The main difference between MBs and C&Cs is that, in the former, polluters may choose either abatement or taxation to fulfil the target, whereas in the latter abatement is assumed to be undertaken by all firms.

The Case Study Area: Upstream Tietê River Basin

Tietê river flows from east to west, with its source near the city of São Paulo at the Serra do Mar, and its mouth at the Rio Grande river 1,025 km downstream. Its sub-basins are the most important rivers in São Paulo state, with many hydroelectric dams and an active navigation. It also supplies fresh water for many municipalities, industries, and agricultural areas, and it is still an important fishery. A major pollution control programme undertaken by the state government is underway in the basin with investments around 1 billion dollars, partly financed by the Interamerican Development Bank.

Cetesb divides Tietê basin into two areas: upstream Tietê and downstream Tietê. As mentioned before, this case study will focus on Upstream Tietê basin.

TABLE 7.8 NUMBER OF PLANTS CONSIDERED IN THE UPSTREAM RIVER BASIN, BY ACTIVITY

Industrial Sector (IBGE) Code	*Activity*	*Number of Plants*
10	Nonmetallic minerals	12
11	Metallurgy	101
12	Mechanical	14
13	Electroelectronic	16
14	Transportation material	29
15	Wood	1
17	Pulp, paper	20
18	Rubber	4
19	Leather, tanneries	5
20	Chemical	64
21	Pharmaceutical	7
22	Soap, candles	6
24	Textile	42
26	Food	51
27	Beverages	16

The Upstream Tietê area covers 32,005 km^2 from the Tietê's source near São Paulo to Barra Bonita dam 592 km downstream, where water quality is classified as "bad." Apart from the Tietê river basin itself, which is divided into three sub-basins, other sub-basins lie in the study area: Billings/Guarapiranga, Capivari, Cotia, Jundiaí, Piracicaba, and Sorocaba. The area covers 121 municipalities that are the most populated and richest regions in the country, and it includes the heavily polluting Metropolitan area and other important industrial zones. Industrial sectors in the region are very diversified and the Cetesb database contains information on 388 plants (53 percent of the total plants registered for the entire state in the database) for fifteen different activities, as shown in table 7.8.

Total residual discharges in the basin reach 11,464 ton of BOD per year. Table 7.9 shows BOD potential and residual discharges and the resulting abatement level, segregated by industrial sector.

A sectoral analysis shows that chemical, food, and textile industries together are responsible for about 76 percent of the potential and residual BOD output (see table 7.8 for a key on sector codes). Other important polluters are transport, with 12 percent of residual BOD emissions, and pulp and paper, with 6 percent. All other sectors have residual outputs smaller than 2 percent of the total. The average

TABLE 7.9 BIOLOGICAL OXYGEN DEMAND EMISSION AND ABATEMENT IN UPSTREAM TIETÊ AREA

Sector	*10*	*11*	*12*	*13*	*14*	*15*	*17*	*18*	*19*	*20*	*21*	*22*	*24*	*26*	*27*	*Total*
Potential (t)	3.6	174.1	46.7	63	1,881.9	0.5	2,076.9	7.5	241.1	11,058.1	7	1,445.7	4,230.7	9,415.9	1,833	—
Residual (t)	1.7	141.8	36.9	57.2	1,346.5	0.3	779	1.6	146.1	3,113.5	2.6	153.4	3,185.4	2,298.5	200.1	—
Abatement (%)	53.5	18.6	21.0	9.1	28.4	30.0	62.5	78.2	39.4	71.8	62.3	89.4	24.7	75.6	89.1	64.7

abatement level is 64.7 percent, ranging from 89.4 percent (textile industries) to 9.1 percent (electroelectronic industries). All major polluters abate above the average.

The Piracicaba river basin has the greatest potential pollution, with 17,203 ton/year because of its important sugarcane alcohol industry, and it is alone responsible for 53 percent of total Upstream Tietê potential pollution of 32,485 ton/year. Other sub-basins with high potential pollution are Tietê Alto Zona Metropolitana (9,501 ton/year), Jundiaí (2,357 ton/year), and Tietê Alto Cabeceiras (1,231 ton/year). All other sub-basins have potential emissions under 1,000 ton/year.

Tietê Alto Zona Metropolitana is responsible for about 53 percent of the residual BOD discharges in the basin. Piracicaba river residual discharges represent 21 percent of the total residual discharge, which is a much lower percentage than that observed for potential pollution because of the BOD removal already practiced by the sugarcane industry.

RESULTS AND CONCLUSIONS

The model was run for the MB case with unit taxes per ton per year and an abatement target from zero to 100 percent. By definition, the C&C results will depend only on abatement targets.

Before discussing MB results, it is worth noting that abatement unit costs vary according to the scale of load abatement. Therefore, special attention should be given to minor value variations. It is possible to estimate large differences in costs or revenues resulting from a slight rise either in tax levels or abatement targets when (1) a higher control step is taken—with significant marginal differences and, consequently, increasing control costs, or (2) no further control is taken—since the next step is much more expensive, but the fiscal revenue will increase.

It should also be noted that the final abatement level in the basin is 64.7 percent but in each plant and sector this level varies around this total average figure. Therefore, when a target is imposed, which is relative to the potential emission, it can be expected that some plants or sectors are already meeting it. For example, as will be seen later in the C&C model, a target of 50 percent on plants will result in a final abatement of 83 percent in the basin.

The model generates results for the typical C&C approach, in which all plants are compelled to abate according to the target set by

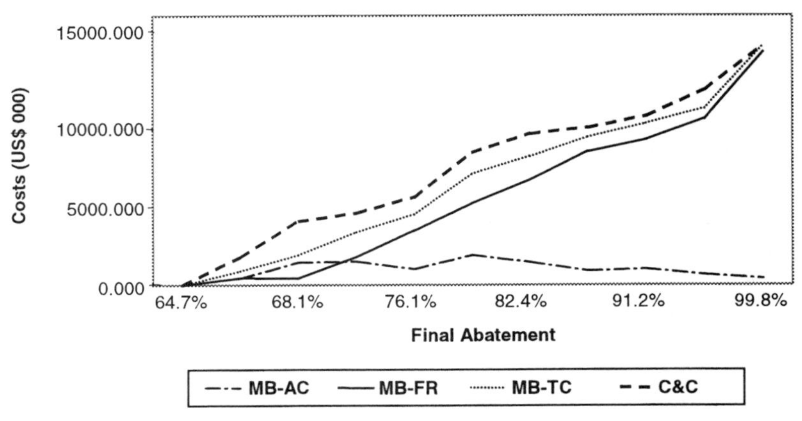

FIGURE 7.2 Isocontrol Curves

the environmental authority, and it also generates estimates for the MB approach, in which plants can choose either to abate or pay tax on emissions below the legal target. The estimates are given for several combinations of tax and target levels and indicate the following results:

1. the final basin abatement (FA) as the percentage of the total potential emission from all plants of the basin which was abated after a tax and/or a target is fixed;
2. the respective abatement control cost (AC), which represents the social costs to achieve a certain level of FA;
3. the resultant fiscal revenue (FR) ; and
4. the total private cost (TC), equivalent to the sum of control cost and fiscal revenue representing the costs accruing to polluters.

Additionally, C&C abatement costs are presented for each FA.

In figure 7.2, some model estimates are presented for a tax level of $5.00 per ton per year combined with different target levels. These combinations led to final abatement results from the current level of 64.7 percent up to almost 100 percent control.[12] Since firms will also choose tax rather than control when the cost of the former is lower, then, in the case of continuous marginal costs, total costs in MBs should not exceed those in C&Cs. As can be seen in figure 7.2, total costs are quite similar in both approaches, with small differences due to the scale effect of the technologies adopted in the study. As mentioned in the previous section, control costs have a scale effect that

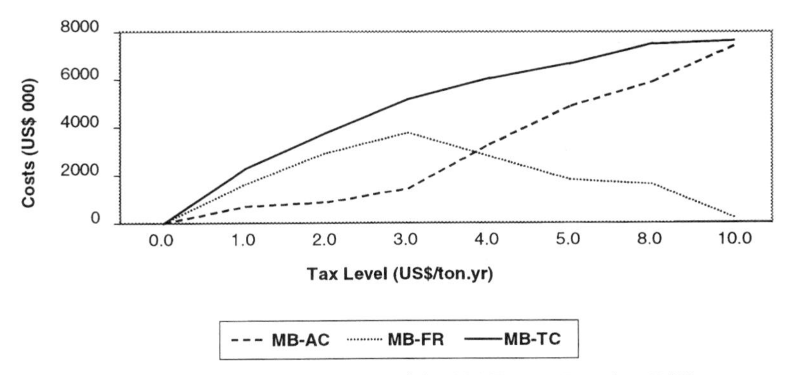

FIGURE 7.3 Market-Based Strategy with 50% Target Level at Different Effluent Tax Rates

does not allow continuous abatement cost. As also expected, MBs always incur lower ACs than those relative to C&Cs—i.e., social costs are lower in MB cases. Fiscal revenue decreases when higher levels of abatement are achieved, since emissions to be levied also decrease.

Figure 7.3 analyzes further the MB results when a fixed target of 50 percent is combined with different tax levels. It can be noted in figure 7.3 that the higher the tax level, the lower is the fiscal revenue because tax costs exceed abatement costs inducing plants to invest in control, as indicated by the increasing control cost curve.

Other possible results of the simulation model can be seen in table 7.10, which shows that the same FA level may be achieved with distinct combinations of tax and target levels. Therefore, one may then construct isoabatement curves using the combination possibilities between target and tax levels. That is, moving along these curves it is always possible to achieve the same FA with different combinations of tax and target levels. In figure 7.4, some isoabatement curves are presented. It is worth noting that the shape of each curve is associated with the technologies assumed in this study.

Table 7.11 presents in detail some points along the isoabatement curve of FA 83 percent.[13] That table shows once again that C&C control costs always exceed those for MBs, which results in higher unit cost for each ton abated. Fiscal revenue decreases when tax level increases. Consequently, the revenue for each abated ton also decreases. Since total costs differ only slightly, as a result of scale effect, the welfare level along the isoabatement curve is the same. That is, the

TABLE 7.10 ESTIMATES OF POSSIBILITIES FOR FINAL CONTROL[a]

Target Level[b]	*Taxation Level (US$/ton/yr)*								
	1	*3*	*5*	*8*	*10*	*20*	*50*	*100*	*C&C*
10%	65.6	67.7	68.0	68.0	68.0	68.6	68.8	68.9	68.9
20%	65.8	68.0	68.1	68.9	68.9	70.4	70.7	70.7	73.8
30%	70.2	71.2	72.5	73.3	73.6	74.4	74.5	74.5	75.2
40%	70.8	73.8	76.1	76.5	76.6	77.2	77.4	77.4	77.5
50%	72.6	74.8	80.4	81.4	82.9	83.2	83.2	83.3	83.3
60%	73.8	80.3	82.4	83.3	83.4	85.2	85.7	85.7	86.8
70%	74.4	82.7	87.2	88.0	88.0	88.1	88.2	88.2	88.2
80%	76.1	89.5	91.2	92.0	92.0	92.1	92.1	92.1	92.1
90%	80.3	92.2	93.7	93.8	94.5	95.4	95.5	95.5	95.6
100%	83.3	97.5	99.8	99.8	99.8	99.8	99.8	99.8	99.8

[a]Percentage of potential emissions from all controlled plants.
[b]Percentage of potential emissions from each taxed plant.

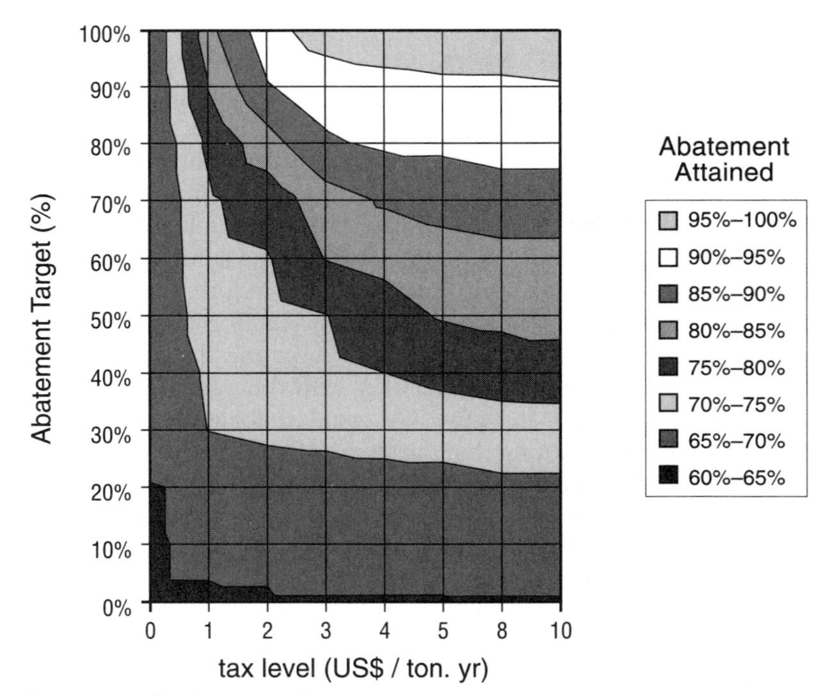

FIGURE 7.4 Isoabatement Curves

TABLE 7.11 ISOCONTROL POINTS WITH 83% OF BOD CONTROLLED

Taxation Levels ($/ton/yr)	*Target Level*[a] *(%)*	*Control Costs ($1000)*	*Fiscal Revenues ($1000)*	*Total Costs ($1000)*	*Cost per Unit Controlled ($1000/ton/yr)*[b]		
					Control	*Tax*	*Total*
1.0	100.0	2,220	5,434	7,654	0.37	0.90	1.27
2.5	74.8	4,731	3,135	7,866	0.79	0.52	1.31
7.0	58.0	7,102	628	7,730	1.19	0.11	1.30
10.0	55.5	7,682	46	7,728	1.27	0.01	1.28
20.0	52.0	7,683	22	7,705	1.28	0.00	1.28
50.0	50.0	7,676	22	7,698	1.28	0.00	1.28
100.0	50.0	7,699	0[c]	7,699	1.28	0.00	1.28
C&C	50.0	7,889	0[c]	7,889	1.31	0.00	1.31

[a]Percentage of total plant emissions that is taxed.
[b]Columns 2, 3, and 4, respectively, divided by the total controlled.
[c]Zero value is the result of rounding.

average total cost for each ton being controlled is equivalent for all options. Therefore, any combination will generate the same welfare change for the society as whole.

However, total and sectoral costs and revenues will largely differ for each policy option adopting a certain combination of tax and target levels. Analyzing the results of table 7.5, it can be seen that a larger target—i.e., a larger tax base—associated with a lower tax level results in lower control costs and higher fiscal revenue. That is the case of the combination $1.00 × 100 percent, which represents the pure MB approach since it requires full abatement of every plant's emission. This case also shows that for that combination of tax and target, society would incur control costs of $2.2 million, whereas a C&C approach, to achieve the same level of abatement, would require expenditures of $7.9 million. That is, in the MB approach, society would have at its disposal the amount of approximately $5.7 million in fiscal revenue to ameliorate the current funding problem in the environmental protection sector.

It is also shown in table 7.11 that a tax increase from $20 to $50 does not raise fiscal revenue since it is not enough to induce further abatement in residual emissions, whereas a much higher tax level, such as the case of $100, will lead to full compliance with the target and result in no revenue at all.[14]

Summing up, the target must be reduced when the tax increases, in

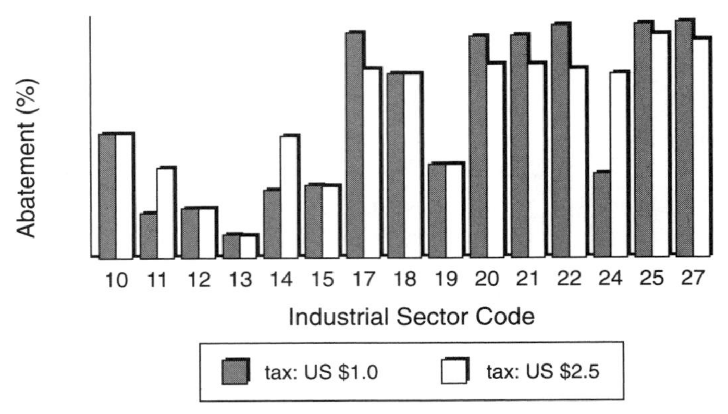

FIGURE 7.5 Isoabatement Curves Comparing Different Target Levels and Tax Rates

order to achieve a certain abatement level. As a target is a proxy for standards, policies seeking fiscal revenues must set lower taxes with high target levels. Since a high tax will make it viable to invest in pollution abatement in those polluting firms with high marginal abatement costs, the size of fiscal revenue is not necessarily linked to the level of taxation but is a result of a combination of tax and target levels.

As shown in figure 7.5, larger targets with lower tax levels alter the sectoral distribution of abatement. In the cases presented in this figure, when a combination of \$1.00 × 100 percent is compared to the equivalent option of \$2.50 × 74.8 percent, sectors 11, 14, and 24 increase abatement to compensate the decrease in others. Consequently, the distribution of costs and revenues will also be different as shown in table 7.12.

Therefore, the decision to introduce a market-based instrument will certainly lead to lower abatement costs than those observed in a command and control approach. However, if a certain level of final abatement is defined, policy makers have to consider environmental standards imposed on individual plants as also being a key parameter to define the level of control and abatement costs in MB policy. That is, tax level is not the unique parameter, particularly if there is a goal to raise revenue in the environmental policy being designed.

TABLE 7.12 SECTORAL COSTS AND INCOME BY POLICY OPTION

	Tax = US$1.00/ton/yr Target = 100%		*Tax = US$2.50/ton/yr Target = 74.8%*	
Sector[a]	*Control Cost*	*Fiscal Revenue*	*Control Cost*	*Fiscal Revenue*
10	0	1.68	0	3.11
11	0	141.75	69.69	178.6
12	0	36.91	0	65.67
13	0	57.2	0	105.73
14	0	1,346.53	793.7	1,238.47
15	0	0.32	0	0.51
17	377.08	99.32	224.86	2.86
18	0	1.64	0	0.41
19	0	146.08	0	217.7
20	1,031.84	733.92	419.86	414.05
21	1.19	0.4	0.74	0.19
22	40.85	23.01	2.27	0
24	137.16	2,755.46	2,843.12	907.18
26	580.21	86.2	363.52	0.35
27	51.26	4	13.67	0

[a]For sector codes, see table 7.8.

APPENDIX

PROBIT MODEL ASSOCIATING WATERBORNE MORTALITY CASES WITH SANITATION SERVICES

A linear model associating social variables and sanitation conditions with z is proposed. This model can than be described in the following way:

$$y = f(z), z = \beta X + \mu,$$

where X is the independent data matrix, y is the observed mortality rates vector, and z is a vector of critical values in the appropriate probability distribution.

The dependent variable (y) in the model is the probability that an individual may die as a result of exposure to waterborne disease. It is here considered that a trial is made for every individual every year. The variable y can assume two values: 1, if this individual dies as a result of these particular diseases, or 0 if anything else happens. Thus y is a qualitative dependent variable; therefore, an unseen variable z is used in the linear model related to the probability of the value of y being 1. By the normal cumulative density function, CDF, one can have a bijection between the probability set and R (the reals). As z can assume any value in R, it can be used in a linear model.

The model then assumes that the social and sanitation conditions produce an unseen value of z. At each year, and for each person, one value is taken from a normal distribution. Only if the value is found to be below z is this person treated as a mortality case.

Therefore, z is the argument of a standardized normal CDF that points to the probability of an individual's dying:

$$P(\text{death}) = \Phi(z) = \int_{-\infty}^{+\infty} \frac{e^{-t^2/2}}{\sqrt{2\pi}} dt.$$

If just one or a few trials are available for each group (or observation), then the observed probability can be 0 or 1 with an associated probit of $-\infty$ or $+\infty$. Such a model can be fitted only by maximum likelihood iteration.

However, since we are considering each person a trial, the number of trials is very high (over 37,000) and in no observation was a zero probability found. This allows the use of ordinary least square (OLS) to estimate z—the probit of the observed mortality rates.

The OLS estimation is heteroscedastic, and the appropriate one would be a weighted least square (WLS), where the ith observation is weighted by $1/var(\mu_i)$. Proofs can be found in Judge et al. (1985:763) and Amemya (1981:1498) that

$$\text{var}\,(\mu_i) = \frac{P_i(1 - P_i)}{\eta_i[\phi(\Phi^{-1}(P_i))]^2},$$

where Φ is the density probability function (DPF) of the normal distribution, P_i is the true possibility in the ith observation, and η_i is the number of trials in i. A feasible WLS, called by Amemya (1981:1491) a "minimum chi-square estimator," can be used, provided a suitable estimate of P_i is available. Both Judge et al. (1985:763) and Amemya (1981:1499) recommend the use of the P_i estimate fitted by least-squares regression without weighting. So, the estimator of P_i considered here is a result of a two-step least-squares regression, where in the first step the weights are estimated by the second one. They were obtained by a Newton-Raphson algorithm made with the SAS PROBIT procedure. Variances and co-variances of the coefficients were corrected by an heterogeneity factor of approximately 45.

NOTES

1. Other diseases, such as those associated with vectors (for example, malaria), were not considered, since their relationship to causes other than water quality are very difficult to segregate. It should also be noted that morbidity cases for the diseases analyzed have a different number of cases and distribution.
2. See, for instance, Martines et al. (1991) for a review of these issues.
3. Provision in rural areas is very low, with only 16 percent of rural population served with treated water, while sewage services are still more pre-

carious. However, it has to be recognized that in terms of negative effects, urban population must be a priority.

4. As will be discussed later, this exercise is an extension of a previous study by Serôa da Motta and Mendes (1993), for which the database was extended from two to nine years and a new mortality data source was obtained to reduce considerably the likelihood of bias.
5. The assumption of the normal distribution being standardized is done so as to have no loss in generality.
6. See, for instance, Greene (1993: 465).
7. See Serôa da Motta and Mendes (1993) for a description of this estimation.
8. For further discussion on these issues of MB application, see Tietenberg (1990), Anderson (1991), Pearce (1991), and Serôa da Motta (1991).
9. No plant was identified individually by name or label.
10. Other options to achieve desired abatement targets are modifications in industrial processes, installation of a community plant for pollutant removal, effluent discharge in public sewerage network, or combinations of all four alternatives. These options were set aside from the model in the interest of simplicity, since marginal costs for these alternative strategies are very difficult to evaluate.
11. Note that, because of data constraints, the environmental target here is based on effluent loads and not on the concentration levels usually employed in environmental policies. However, that distinction will not change the results.
12. Organic matter control is possible up to 99.7 percent with the technologies being adopted in this study.
13. In fact, the FA value is between 83.0 and 83.3 percent, since a precise percentage is not possible because of the combination of technologies.
14. In the case of other pollutants, for example heavy metals, a level of almost 100 percent abatement is not technologically possible. Therefore, fiscal revenue will not be nil but constant if target level is over the highest abatement level achievable.

8

Valuing Social Sustainability: Environmental Recuperation on *Favela* Hillsides in Rio de Janeiro

PETER H. MAY

Progress toward the macro goals of global and national sustainability can be made only if there is equivalent progress toward building local and regional alternatives to current patterns of resource degradation and human squalor (ICED 1992). Successes from efforts in this direction have come through societal learning and empowerment, which enable communities to define and act on problems through a discursive coevolutionary process (Norgaard 1994). The experience gleaned from such endeavors offers knowledge applicable toward a range of problems whose resolution nevertheless must rely on each locale's unique social and political context. To balance the uncertain facts and competing values paradigmatic in socioenvironmental conflict, "extended peer review" (Funtowicz and Ravetz 1994) must reach beyond the realm of science itself to include the communities that are the subjects of policy and intervention. These precepts are particularly valid in the developing world, where social inequities are brutally juxtaposed with environmental problems (Environment and Development Commission for Latin America and the Caribbean 1990).

In evaluating one case of public intervention to allay socioenvironmental conflict in the setting of a developing country, this study reflects a concern with what I have here called social sustainability. Such an approach considers not only the benefits and costs of investments, but also the social and political forces and the local perceptions that condition efforts to reconcile environment and equity in developing countries.

URBAN POVERTY AND RIO'S *FAVELAS*

Coming to grips with poverty in the developing world means confronting urban problems. Urbanization has assumed significant proportions in many developing nations, particularly in Latin America's southern cone, where on average 75 percent of the population now lives in cities (World Bank 1992). Rural households whose labor has become superfluous to an increasingly mechanized agribusiness complex have been drawn to the cities by the possibility of access to schools and health care, and by the magnetism of industrial and service employment. But many soon discover not only that these opportunities remain elusive, but also that urban land is scarce and expensive, forcing them to seek unoccupied spaces that are invariably subject to environmental hazards and that are rarely equipped with the services that initially attracted them.

The city of Rio de Janeiro is no exception to the pattern of explosive urban growth experienced by most developing nations. Compressed in a narrow strip between the seashore and the steep granite slopes of the Atlantic coastal range, Rio's pattern of spatial occupation exposes a stark social stratification that contrasts sharply with its physical beauty. Those better-off live on the urbanized lowlands, well equipped with infrastructure and services. The poor, on the other hand, live primarily in flood-prone shanty towns or in hillside slums known as *favelas*—a generic term that refers to improvised housing on lands lacking well-defined property rights or public services. In 1993, there were 570 *favelas* scattered throughout the Rio de Janeiro metropolitan region, serving as home to about 1 million people (over 10 percent of the metropolitan region's total inhabitants). Inhospitable terrain and immoderate densities combine with poverty and violence to make these settlements the blight of rapid urbanization.

Rio's historical pattern of development resulted in an obliteration of natural forest cover, which declined from an estimated 97 percent of state land area in the era of colonization to 12 percent by the end of the 1980s (SOS Mata Atlântica 1992). Efforts to curtail detrimental downstream effects of deforestation occurred in two phases. In the eighteenth century, coffee cultivation blanketing the countryside had laid waste the steep forested slopes surrounding the city, causing lowland floods and landslides. The Imperial government of the mid-nineteenth century successfully reforested and protected slopes in the critical Tijuca Forest watersheds, assuring natural forest regeneration.[1]

In the mid-twentieth century, a second rapid urban growth phase was accompanied by the decimation of remaining natural cover on unprotected hillsides. Widespread construction of access roads and housing in *favelas* led to increased surface runoff and sedimentation. Despite engineering solutions that channeled streams into culverts and canals, the city again became prone to frequent flooding, a problem further exacerbated by inadequate sewerage and solid waste collection. Areas that had been forested succumbed over time to degraded pastures of the aggressive elephant grass (*Panicum maximum*), which easily caught fire in the dry season. Natural regeneration became nearly impossible because of the subsequent high mortality among tree seedlings that struggled to gain a foothold on the steep slopes. The failure to define limits to settlement expansion or to encourage natural forest regeneration soon brought calamity, threatening lives and property.

Up to the 1950s, public authorities responded to these land-use conflicts by razing slums and erecting cement slope buttresses to protect the homes below against landslide. Yet it became clear that the forced relocation of slumdwellers simply transplanted urban problems to other parts of the city. Expanded urbanization made the housing deficit ever more acute. Public housing projects were decried by a growing squatter movement as destructive to community values. The participants in such movements portrayed such projects as counterproductive, in that those removed from the *favelas* were rarely able to afford even subsidized public housing payments, forcing them to swell squatter refugia elsewhere in the metropolis. By the 1980s, as a result, municipal governments began to accept that the *favelas* should be treated as an integral and vital part of the urban fabric, and they vowed to allocate resources toward making them more livable (Kreimer et al. 1993).

Despite this new orientation on the part of public authorities, hillside communities, which count for half of Rio's *favelas,* remained vulnerable to natural disasters. In 1988 a brutal storm left hundreds of hillside homes crushed by rockslides; many died and thousands were left homeless (Munasinghe et al. 1990).[2] As is often the case, crisis finally led to action on the part of the municipal government. This study evaluates one of the responses to socioenvironmental conflict that emerged in Rio in the wake of that storm.

COMMUNITY REFORESTATION PROJECT

Soon after the immediate devastation of the 1988 storm had been swept up, at great cost to local governments, a team of foresters and social service workers in the city government of Rio de Janeiro conceived the Community Reforestation Project (*O Programa de Preservação e Reflorestamento em Regiões de Baixa Renda—Projeto Mutirão-Reflorestamento,* hereafter called the Project). The Project aims to recover forest cover on degraded hillsides located above *favela* communities, at a low cost, with the aim to induce greater groundwater infiltration, reduce runoff, and enhance slope stability. Sites selected for this effort were defined according to the following criteria: (1) they consist of deforested hillsides with steep slopes, subject to landslides or rockfall; (2) the areas are prone to flooding, sedimentation of rivers, or drainage channels; (3) the areas have become subject to patterns of human occupation that constitute a risk to residents and neighboring areas; and (4) the settlements evidence some measure of community organization.

This Project thus has a clear social connotation, not only because of its target population and sites, but also because it involves local community members in its implementation. Reforestation and drainage works are carried out by local residents named by community associations and remunerated by the city government for their role in the Project. Aiming to reduce costs associated with natural disasters, the Project would be of benefit to the public purse, to populations located in the area of influence of the affected hillsides, and to those who live on the hillsides themselves. Between its first pilot efforts in 1988 and the time of this study in 1996, the municipal government expanded the range of Project actions into over twenty-seven *favelas* throughout the metropolitan area, there having been an estimated 100,000 direct and indirect beneficiaries of the 400 hectares (ha) of steep terrain reforested (UNDP/IBAM 1993).

The approach toward environmental recuperation involved the planting of native species in patterns that would mimic natural succession of tropical forests. Species were selected for their rapidity of establishment and root spread to maximize soil protection and retention, as well as for their rusticity, which would ensure survival on the hazardous hillside environment. These include nitrogen-fixing legumes and, typically, an assortment of fruit trees to further stimulate community interest in protecting the reforested area.[3] Periodic campaigns accompanied this effort, through fliers, films in public schools, and "consciousness-raising visits to each residence in the *favela* in an attempt to show the importance of the forest to the hillside."

Such a procedure, however, was not always possible to adopt as time went on. This occurred as a result of a lack of both financial and human resources, limiting the work that could be accomplished

(A)

FIGURE 8.1 Morro São José Operário: (A) before Project, (B) after Project.

to planting and maintenance of seedlings. In reality, the local government agency lacked the support necessary to augment the scale of the project, as had originally been forecast, because of political discontinuities that occurred in the transfer of municipal administration after the 1992 elections, and because of a setting that led to the Project being perceived at some times as a high priority, and at others as marginal for engendering political capital.

In examining Project actions in one beneficiary community, the following case study assesses features of these actions and the roles of involved stakeholders, for the purpose of considering the potential contribution of its design and outcomes to socially sustainable cities in the developing world.

COMMUNITY PROFILE

The municipal government initiated its actions under the Community Reforestation Project through pilot investments in the *favela* Morro São José Operário in the Jacarepaguá district in Rio de Janeiro, whose residents and their homes had been severely harmed by the storms of 1988. Degraded, deforested, with unstable rock outcroppings and steep slopes with an average 46 percent incline, the 1,000-foot summit of São José Operário had also been the site of a clandestine granite quarry, which further placed the population at risk.

(B) 

Access roads were cut and soils disturbed by stone cutting, dynamiting, and transport, which provoked gullying and sedimentation. The community situated on this hillside began to grow in the 1950s; most of today's 4,500 households have lived there for more than ten years.

Project interventions were focused on a microwatershed covering 32 ha; hundreds of residents' shanties densely blanket the lower watershed (figure 8.1A). Many of these homes lack foundations and are positioned on risky sites: because of the hillside's steep incline, they are constantly threatened by rolling boulders and landslides. Figure 8.2 shows the topographical relief of the project area, indicating the location of major unstable rock outcroppings and their line of fall.

On this hillside, the Project aimed to reforest unstable slopes, thus (1) reducing the erosion and sedimentation that obstructed drainage channels and had caused floods on residential streets lying at the base of the hill, and (2) averting accidents from rockfall. As related objectives, it was hoped that the watershed would be preserved and replenished, and that conflicting land uses would be avoided by restricting the expansion of the community in the risk-prone area. Figure 8.1B shows the effect of Project-sponsored reforestation on the visual character of the community.

According to results of a survey conducted for this study, São José Operário is socially stratified according to the location of shacks on

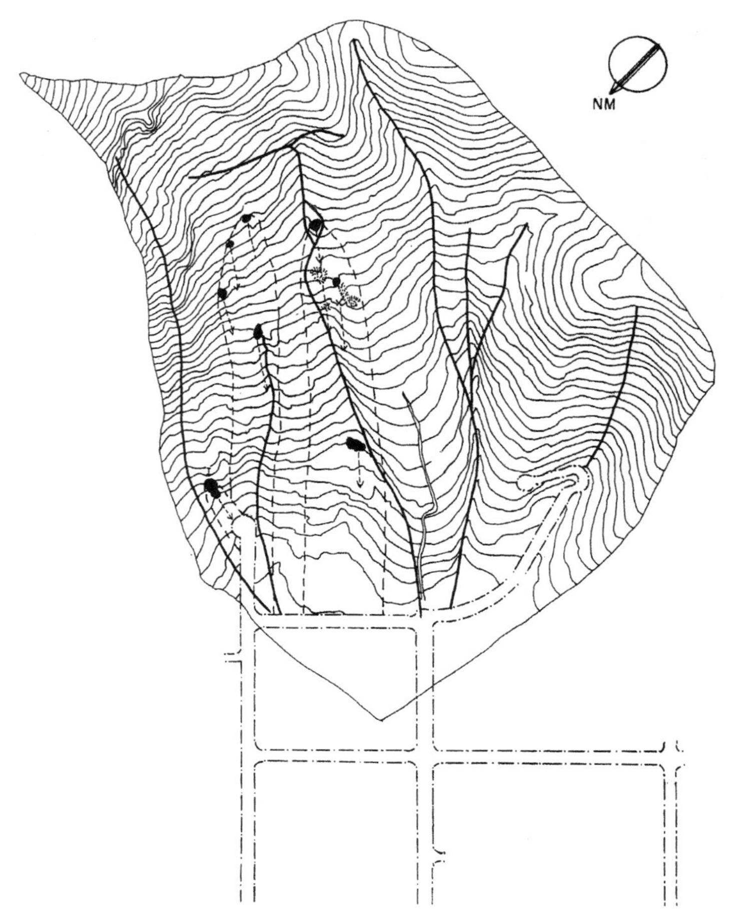

Figure 8.2 Topographical Features of Morro São José Operário

the hillside. To buy a two- or three-room shack, residents pay on average about $800,[4] a value that can double for those located "from the church on down," the community's most accessible and hence most coveted sites. Those homes situated nearer the hilltop are cheaper than the average price, not only because of their difficult access, but also because they lack the most basic sanitation facilities. These families must individually pipe their own water supplies from springs on the hilltop, and these springs—because of the lack of watershed protection—had been prone to dry up, making it necessary to carry water to the home from a more distant standpipe. The community, acting in collaboration with the municipal water authority,

Table 8.1 Sources of Degradation, Environmental Effects, Receptors, and Municipal Interventions in São José Operário, Rio de Janeiro, Brazil

Origin	*Effects*	*Receptors*	*Interventions*
Inadequate hillside occupation; granite mining	Deforestation/ burning/grass invasion; erosion/ sedimentation; lowland flooding; risk of rockfall/ mudslide	Local population; ecosystem/ watershed; collective infrastructure; homes/ improvements; downstream community	Reforestation with native fast-growing species; mining prohibition; containment of hillsides and rock outcroppings; drainage works
Lack of basic infrastructure	Health costs; high cost of service provision (water supply, waste disposal)	Local population, particularly children (high diarrhoea incidence)	Storm sewerage works; paving; trash removal/ recycling

had built and maintained a pumping system distributing water to remaining *favela* dwellings from a tank halfway up the slope.

Residents' incomes of about $110 per household per month for the average family of five is barely enough to purchase a basic food basket consisting primarily of rice, beans, and cassava flour. Household heads are employed in civil construction or in odd jobs, while women usually find work as domestic servants—often far from the community. There is one primary school on the hillside, but residents must descend to the neighboring community to obtain health care.

Table 8.1 summarizes the principal factors that had caused degradation in the local ecosystem, and the negative impacts of these factors on both ecosystem and the resident and downstream communities. It also describes those interventions carried out by the community with technical and financial support to the Project from the municipal government in an effort to reverse these problems.

ENVIRONMENTAL VALUATION

The valuation procedure applied by the study first characterized the extent to which the Project averted risks faced by local residents, and

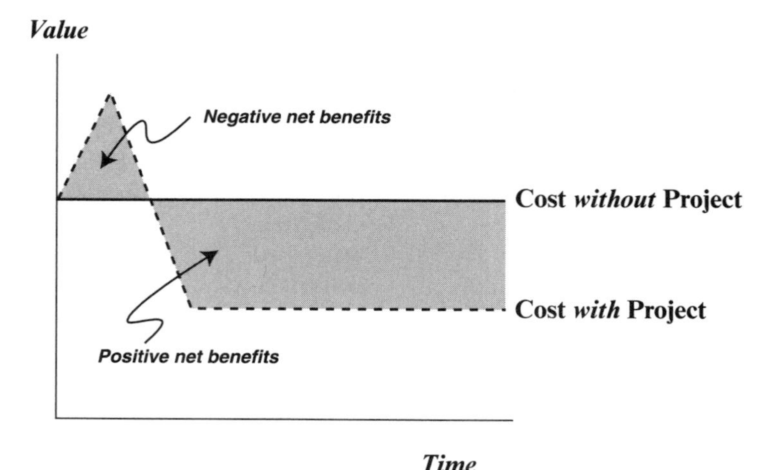

Figure 8.3 Incremental Net Socioenvironmental Benefits of Project Actions

then it quantified these benefits in monetary terms. This represents an incremental net benefits approach (Squire and Van der Tak 1975) assessing social costs associated with environmental degradation before and after the project was implemented. This approach treats a reduction in costs as a benefit from the intervention, net of reforestation investment, and associated direct costs. This analysis makes use of the following expression of net social benefits:

$$\sum B_{it} = \sum[(C_b - C_a) - C_p]_t,$$

where

C_a = environmental costs after recuperation;
C_b = costs anticipated prior to investment;
C_p = costs of Project implementation and maintenance;
B_i = incremental benefit from investment (all assessed at period t).

Figure 8.3 illustrates this concept graphically, reflecting the temporal behavior of the benefit flows, showing that during an initial phase (public works and planting), the net benefits can be anticipated to be negative. Only once the vegetation is fully established can an improvement be seen in the sedimentation and other damages that would have occurred without the investment. Full realization of these gains is only obtained after a number of years have passed.

The principal natural capital investments made by the project involved establishment of over eighty species of native and exotic fast-growing trees, as well as the maintenance of the reforested areas and of drainage works installed. Hillside reforestation—including wages, tools, inputs, and transport costs—totaled $2,308 per hectare reforested and $4,611/ha/yr during the three-year maintenance phase, for an overall gross cost of $16,143/ha. These relatively high maintenance costs can be explained by continued payment of salaries to the local six-person reforestation crew over the latter period, to ensure adequate seedling protection and community engagement in the process. The estimated cost of construction of the 1,000 m of drainage works installed was $17,311, undertaken during the first project year. Overall costs of the project, involving reforestation of 9.8 ha and installation of drainage channels, were thus $175,412, spread over a four-year period.

Benefits identification reflects the perceptions of municipal employees and community members, as well as residents of the adjoining bottomland streets. Consistent with the aims of the Project, these benefits include the reduction of risk from flooding and rockslides, reduced costs of clearing debris from stormwater channels, adjustment in real estate values, assurance of reliable springwater flow, and expanded space available for community recreation. The procedure for estimation of monetary benefits is sensitive to the fact that sediment reduction and related risk could not be alleviated until tree cover had fully taken hold, and that environmental risks will never be completely eliminated as long as the community occupies this precarious slope. Thus the analysis estimated that no benefits accrued until after the initial four-year planting, replanting and maintenance period—the situation depicted in figure 8.1B. However, to simplify assessment (and because of the unavailability, at this writing, of soil-loss data being gathered through ongoing research at the Project site), an average based on the universal soil loss equation (USLE)[5] was calculated rather than relying on uncertain projection of progressively declining erosion rates.

Benefits estimation took account of the following components:

- Sediment reduction: Prior to the Project, based on USLE estimates, an average of 40 t of sediment and solid wastes were deposited to streets below the hillside during torrential storm events, approximately once each year, making necessary their removal in dump trucks by the municipal government. These loads were reduced by the Project to about 14 t/yr, thus cutting municipal expenditures by $873/yr.

- Reduced flood loss: Periodically, about once in every ten years, a catastrophic storm flooded a 23 ha area lying at the foot of the hillside, damaging the 2,000 homes situated there. These property damages were reduced by the reforestation project by $100 per household, for a total benefit of $200,000. In addition, working members of these households would no longer lose work days as a result of the difficulties caused by these catastrophic floods, an incremental benefit valued at $14,000.
- Reduced rockfall risk: Fifty homes lay in the path of potential rockfall, mapped from aerial photographs, constituting risk of complete destruction and associated danger to life and property. The Project averted a great deal of this risk with reforestation complemented by minor civil works to secure unstable boulders. These investments were estimated to bring reduced property losses of $50,000, which were projected at intervals of five years, beginning ten years after Project initiation.
- Property revaluation: Reduced sedimentation and a safer living environment have resulted in a 20 percent improvement in property values, principally in the area "from the church on down," occupied by about 500 homes, for a total value of $100,000, accruing on a one-time basis on attainment of young forest cover five years after project initiation.
- Springflow stabilization: Some forty homes located on the slope above the community's water distribution tank are now assured a more even flow of water from springs on the hilltop, because of greater infiltration in reforested areas (calculated based on differential groundwater recharge rates for altered vegetation cover conditions on the soils and slopes in question). This reduced the frequency with which residents had to descend steep paths to obtain tapwater, or had to pay others to deliver it, a benefit valued on the basis of municipal water service charges on a volume basis at $12,348/yr beginning in the fifth year after Project initiation.
- Recreation site provision: As a result of the provision of shade and fruit trees, the community now possesses a recreation area, which reduces costs incurred previously when residents visited other open areas for leisure pursuits, valued on the basis of bus transport costs for a monthly visit by 500 families whose four members were directly benefited by the Project, totaling $16,800/yr beginning five years after project initiation.

TABLE 8.2 PRESENT VALUE OF DIRECT COSTS AND BENEFITS OF HILLSIDE REFORESTATION IN SÃO JOSÉ OPERÁRIO, RIO DE JANEIRO, BRAZIL

Costs and Benefits[a]	*Present Value*
Costs	
Reforestation establishment/maintenance	$117,098
Drainage works	$15,456
Subtotal	$132,554
Benefits	
Sediment reduction	$3,869
Reduced flood loss	$91,087
Eliminated rockfall risk	$30,417
Housing revaluation	$56,743
Springwater stabilization	$54,728
Recreation site provision	$74,459
Subtotal	$311,302
Net benefits	$178,748
Benefit-to-cost ratio	2.35

[a]Costs and benefits in current U.S. dollars discounted over 20 years at 12 percent per year. See text for description of methods used to estimate costs and benefits.

The results of this simple benefit–cost analysis, summarized in table 8.2, suggest that the benefits of environmental recuperation, despite a conservative estimation approach, were considerably greater than their associated costs. Using a relatively high real cost of capital of 12 percent commonly applied as a cutoff rate by multilateral financial institutions, the discounted benefit stream attains a value that is more than double that of the initial public investment. These results argue that the benefits the community obtained from these investments more than justify the public expenditures involved, when viewed from the limited perspective of public finance.

EQUITY EFFECTS

With regard to distributive issues, as the funds applied toward investments in São José Operário were public in origin, the incidence of revenues between social groups and firms is of concern. Since municipal revenues in Brazil are derived primarily from value-added taxation (ICMS) and from property taxes (IPTU),the tendency is for their incidence to be regressive in relation to taxpayers' income. This is true because the marginal tax rate for consumer goods is not differentiated between consumers, and property tax levies do not increase with property values. In this sense, the directing of benefits toward those social groups having least purchasing power might compensate partially for the regressive origin of investment funds. Furthermore, since *favela* dwellers do not pay property taxes, benefits to them represent a net transfer out of the share of public revenues derived from that source.

However, of the benefits assessed, a substantial share went not to *favela* residents themselves but to members of the community who reside on the streets at the base of the hillside. The improved value of property, the protection of material possessions, and the gained worktime that would have been lost as a result of flooding were of benefit more to these households than to those who reside on the hillside. Since the former have greater incomes, this result characterizes an investment whose benefits are distributed inequitably. One compensatory factor in this sense is the fact that the works were concluded through remunerated community self-help, and the majority of investment "costs" reverted directly to the community in the form of wages to individuals who reside in São José Operário.

Some community members were opposed to the proscription of clandestine granite quarrying, arguing that this enterprise stimulated the local economy and provided investments in community facilities. The granite miners had purportedly made financial contributions to the local residents' association, although they employed no community members directly.

In terms of the mining enterprise's social sustainability, the retention of such economic activity may be desirable if it is possible thereby to condition this activity on investments to reverse environmental degradation and provide community members with gainful employment. However, as a clandestine operation, the miners paid no taxes on resource depletion or granite sales that could be targeted toward mined land restoration. As a wildcat operation, the slightest hint of an increase in costs would lead to their pulling up stakes. There was hence no legal recourse or financial remedy available to the community to avert the environmental hazards caused by clandestine mining, short of its prohibition and public investment in mined land recuperation.

COST EFFECTIVENESS

In addition to evaluating incremental benefits, the study assessed the cost effectiveness of the São José Operário Project, comparing the costs of environmental recuperation with alternative measures that could have been used to meet these objectives. These alternatives included (1) the use of cement slope retention to secure against rockfall, (2) reforestation carried out by third-party firms, and (3) the complete relocation of slumdwellers and prohibition of hillside occupation.

Civil works to retain slopes, which would cost substantially more than reforestation, nevertheless would represent only a partial solution. The full range of additional benefits generated by the Community Reforestation approach, such as sediment reduction, shade, and recreational opportunities, would not be provided through an engineering approach. However, by complementing minor civil works with reforestation, the local government reinforced the benefits derived from the former investments.

The second option, reforestation by third-party firms contracted by the municipal government, besides being more expensive, has failed generally to be effective in meeting objectives in Brazilian cities. This has been true because of their inadequate attention to the protection and maintenance of trees once planted (Pastuk 1995). To be effective, reforestation on *favela* hillsides requires that local communities commit themselves to manage reforested areas over the long term. Given the close involvement of the community in selecting the Project workforce and locally responsible parties, maintenance was ensured during the critical initial growth phase. It is hoped that, once fruit trees and shade-offering species grow to maturity, their utility to residents will ensure their protection in the future.

As for the ultimate solution contemplated, relocation would not only represent a major financial burden, but it has also become politically unacceptable. As described, communities that had disintegrated in the wake of prior actions to remove squatter settlements successfully challenged this approach. *Favelas* are now treated as a legitimate part of the urban fabric. The slum clearance option simply no longer exists.

SOCIOPOLITICAL ASSESSMENT

The concept of social sustainability implies that people in general should have their basic needs satisfied not only through a fair income distribution and an equal access to services and facilities provided by the State, but also through an equal opportunity to participate in the political arena as subjects of public policy design and implementation. For this to occur, the perceived needs and concerns of the poor must first be treated as legitimate.

Thus, in the framework of social sustainability adopted for this study, options for environmental protection investment should also be considered in the light of the perceptions of stakeholder groups regarding the relative importance they attribute to these options among concerns affecting their cultural and natural environment.

In the São José Operário Community Reforestation Project, unfortunately, this process did not fully occur. Based on in-depth follow-up interviews, the postulated benefits were, in fact, perceived by only part of the community. True enough, no further losses were registered in homes or improvements after the works were implemented and many springs had indeed began to flow more copiously and with greater regularity, benefiting the residents located near the reforested area near the hilltop. On the other hand, the homes in this area did not increase in value as had initially been imagined. For this to occur, it would have been essential for the residents to have access to basic sanitation services and an easier climb to their homes. Indeed, the works resulted in improved valuation to the shacks situated lower down the hillside, where such services and ease of access already existed.

With regard to the homes located at the base of the hill, reforestation and drainage investments reduced losses that had previously arisen as a result of flooding and sediment deposits, but these continued with lesser intensity because of the solid wastes accumulated on the hillside, which continued to obstruct stormwater drains when carried down the slope by torrential rains. Even for the downstream community, therefore, the benefit was perceived as only partial in scope.

The Project itself was valued favorably, by community members interviewed, only to the extent that it was accompanied by other actions they considered to be local priorities, such as drainage and, more recently, the tying-in of sewerage connections to stormwater drains. In fact, water, sewerage, and trash collection are the items always mentioned when the community made its own accounting. When the community was solicited to indicate the most urgent problems facing them, only 10 percent of its members mentioned reforestation as an issue, while 90 percent mentioned lack of basic sanitation services. It can thus be concluded that a reforestation project would have had greater success, from the *favela* community's point of view, had it been accomplished in close connection with other projects that its members defined as priorities.

Through this analysis, it became clear that the actions perceived as most effective by community members are those that are carried out in the most integrated fashion. Such an integrated approach, however, would contradict the political strategy or logic at play, a logic based on traditional power structures of clientelism and the "ideology of favors." This ensures that the more widely dispersed and numerous are public actions, the more those served will feel they have benefited in some way, bringing more votes and political adhesion for the patron figure. Thus it may not be all that important if these actions have real positive socioenvironmental repercussions but, rather, that some attention has been dedicated by public figures to community problems in a widespread, visible way.

As a further hypothesis, it was initially suggested in this study that as a result of community members' involvement in the Project, greater local organization could ensue, thus increasing local capacity to mobilize around priority works in future. This hypothesis cannot be considered confirmed, however, for two reasons. First, although the municipality hired indigent local residents who had been selected by the residents' association to act as remunerated workers on the Project, broader community involvement was not guaranteed by this strategy alone. Other mechanisms for participation were needed. The environmental education effort was more symbolic than a continuing and integrated aspect of local government engagement, and it was cut back when other political priorities emerged.

Furthermore, São José Operário, like most *favelas* in major Brazilian cities, faced serious difficulties throughout the Project's implementation as a result of local interference from drug traffickers. This led to a rampant air of fear and insecurity among residents, now no longer arising from risk of accidents with natural causes but rather from violence. Warily, they keep the "law of silence," and all problems seem of little importance when confronted with the threat of death from the crossfire to which they are constantly exposed. As can be imagined, further community mobilization that could have been spurred by involvement in environmental recuperation has been stymied as a response to this new societal risk.

It is important to emphasize another aspect that came to our attention through this study: Few members of the community understood the nature of the government project; most did not know they were living under a constant state of risk and showed surprise when this factor was mentioned. This perception illustrates the evident lack of information *favela* residents possess about the cumulative and sporadic risks they face as a result of their settlement conditions. Some even declared that it was a pity that the squatter housing area could not continue to expand as they had wished, because the more visible the *favela,* the more obvious its problems, and hence the greater flow of public funds that might eventually ensue. Reforestation for these people created a false image of prosperity in the eyes of the outside world.

CONCLUSIONS

Sustainable cities represent a significant challenge to the application of the principles of ecological economics to developing-country problems. Unequal access to land because of unfair distribution of income and power forces poor urban migrants to occupy fragile and inhospitable sites, subjecting themselves and others to serious environmental and health risks, a problem common in Rio de Janeiro.

This study has evaluated a municipal project for reforestation of hillside squatter settlements (*favelas*) in the Rio de Janeiro metropolitan region that aims to partially avert such risks. The evaluation took into account a project's social sustainability, considering not only the benefits and costs of environmental recuperation, but also the sociopolitical forces and local perceptions that condition efforts to reconcile environment and equity.

This analysis has revealed a number of important themes that permeate investigation of socioenvironmental conflict in developing societies. On one hand, the role of local stakeholder engagement in defining priority investments has become an evident necessity in the light of innumerable cases of misguided public policies and projects. On the other hand, the evident contradiction between locally perceived values and sheer physical risk highlight the need for an interactive process of project planning that includes initially empowering residents to become more capable of diagnosing the environmental features and problems they face daily. Although reforestation partially ameliorated environmental risks faced by community members in São José Operário, their own perception of the benefits brought by this project was colored by their recognition that a false image of prosperity had been engendered by arborization of denuded slopes—in frontal contradiction to their incessant struggle with poverty. There is a great need therefore to complement environmental recuperation with social empowerment, even while immediate sources of risk were in the process of alleviation. Nonformal environmental education must become a tool toward mobilization of collective action in the economic sphere.

A second related issue is that of the mechanisms for contribution by poverty-stricken communities to local investments aimed to safeguard environmental values. On the one hand, an active contribution is considered to engender commitment to protection and maintenance, crucial to the recuperation of degraded landscapes, particularly on *favela* hillsides vulnerable to wildfires and vandalism. On the other, the poor are seldom able to divert time they could be using to secure basic economic necessities toward sweat equity for community needs. The solution applied in the Community Reforestation Project of remunerating workers selected by the residents' association from among local indigent households may be perceived as paternalistic, but the return of most reforestation costs to the community in the form of employment surely helped to guarantee project success. Nevertheless, such employment must be seen not as an end in itself, but as a means to stimulate community participation and greater mobilization for needed services and investments.

Finally, the paradox of desirably integrated investment versus politically motivated dispersion is a ubiquitous tension not easily reconciled with demands for efficient public resource allocation. The importance of coordination among complementary institutions in unifying efforts to solve complex urban problems has been reinforced by this study.

As has become evident in the preceding analysis, the assessment of social sustainability is an essential complement to economic valuation of environmental investments in developing nations. Insofar as it reveals both the potential and the fragility of traditional cost–benefit analysis, further opportunities are needed to define concepts and methods that can make ecological economic analysis pertinent to decision making in the transition toward sustainable cities.

NOTES

1. The Tijuca Forest is now considered a marvel of urban parkland, which envelopes the sprawling metropolis of Rio de Janeiro, adding to the city's tropical exuberance. Drummond (1988) offers a chronicle of the massive reforestation effort using native species, undertaken over more than a decade by a handful of slaves under the direction of the intrepid Major Archer.
2. In the rainfall of 184 mm registered on February 3, 1988, alone, which affected principally the cities of Petrópolis and Rio de Janeiro, approximately 300 deaths were recorded, an additional 1,000 people were wounded, and losses totaled $1 billion, including the destruction of 4,000 homes.

3. A partial listing of the eighty species used includes sabiá (*Mimosa caesalpiniaefolia*), leucena (*Leucena leucocephalla*), sombreiro (*Clitoria fairchildiana*), acácia (*Acacia auriculiformis*), albizia (*Albizia lebbeck*), ingás (*Inga* spp.), angicos (*Anadenanthera macrocarpa* and *A. colubrina*), jacaré (*Piptadenia gonoacantha*), aroeira (*Schinus theribintifolius*), guapuruvu (*Schizolobium parahyba*), goiabeira (*Psidium guajava*), jamelão (*Syzygium jambolana*), nespereira (*Eriobothria japonica*) and mangueira (*Mangifera indica*). Further details regarding successional principles that guide the Project are reported in Pastuk (1995).
4. Values used in this study refer to current nominal prices converted to U.S. dollars at the tourism rate, an approximation for the border exchange rate. This conversion does not represent purchasing power equivalent, which in Brazil has fluctuated considerably as a result of monetary policies. At the time the study was conducted, purchasing power parity was considerably higher than exchange rate equivalency. Since the Plano Real in 1994, however, purchasing power has declined dramatically. Inflation stabilization achieved by this plan has nonetheless been popular insofar as real earnings of low-income groups improved.
5. The universal soil loss equation has been defined as follows:

$$A = R \times K \times LS \times C \times P,$$

where

A = average annual soil loss in t/ha
R = rainfall erosivity factor
K = soil erosivity factor
LS = inclination factor
C = soil preparation and ground cover factor
P = conservation practices factor

References

ABRACAVE. 1996. *Anuário Estatístico.* Belo Horizonte: ABRACAVE.

Adaime, R. R. 1985. Produção do bosque de mangue da Gamboa Nobrega (Cananéia, 25° Lat. S, Brasil). Doctoral thesis, Universidade de São Paulo, Instituto Oceanográfico.

Aguiar, T. C. 1991. O papel do poder público municipal: Desafios na criação de políticas para o desenvolvimento integral e harmônico da área rural. *Revista de Administração Municipal* 38(200):49–57.

Alberini, A. 1995. Optimal designs for discrete choice contingent valuation surveys: Single-bound, double-bound, and bivariate models. *Journal of Environmental Economics and Management* 28:287–306.

Albuquerque, F., C. Veloso, M. Duarte, and O. Kato. 1989. *Pimenta do reino. Recomendações basicas para seu cultivo.* Belém, Pará: EMBRAPA-UEPAE.

Alexander, A., J. List, M. Margolis, and R. D'arge. 1996. Alternative methods of valuing global ecosystem services. Unpublished manuscript.

Almeida, O. T. de, and C. Uhl. 1995. Developing a quantitative framework for sustainable resource-use planning in the Brazilian Amazon. *World Development* 23:1745–64.

Amemya, T. 1981. Qualitative response models: A survey. *Journal of Economic Literature* 19:1483–536.

Anderson, A. B. 1990. *Alternatives to Deforestation: Steps toward Sustainable Use of the Amazon Rainforest.* New York: Columbia University Press.

Anderson, D. 1991. An economic perspective on management in the public sector. In: Eeröcal, D., ed. *Environmental Management in Developing Countries.* Paris: OECD.

Bateman, I., I. Langford, K. Turner, K. Willis, and G. Garrod. 1995. Elicitation and truncation effects in contingent valuation studies. *Ecological Economics* 12:161–79.

Batie, S. S., and L. A. Shabman. 1978. Economic value of natural coastal wetlands: A critique. *Coastal Zone Management Journal* 4(3):231–47.

Barbier, E. B. 1991. The Economic Value of Ecosystems: 1. Tropical Wetlands. London: London Environmental Economics Center, Gatekeeper series 89–102.

Barreto, P., C. Uhl, and J. Yared. 1993. O potencial de produção sustentável de madeira em Paragominas-PA, na Amazônia Oriental: Considerações ecológicas e econômicas. *Anais do 7° Congresso Florestal Brasileiro* 1:387–92. São Paulo: Sociedade Brasileira de Silvicultura/Sociedade Brasileira de Engenheiros Florestais.

Barros, N. F., and R. F. Novais, eds. 1990. *Relação Solo-Eucalipto.* Viçosa, Minas Gerais, Brazil: Folha de Viçosa.

Bartelmus, P. 1996. Green accounting for sustainable development. In: May, P., and R. S. da Motta, eds. *Pricing the Planet; Economic Analysis for Sustainable Development,* 180–96. New York: Columbia University Press.

Bartelmus, P., E. Lutz, and S. Schweinfest. 1992. Integrated Environmental and Economic Accounting; A Case Study for Papua New Guinea. Washington, DC: The World Bank Environment Department.

Bartelmus, P., and J. van Tongeren. 1994. Environmental Accounting: An Operational Perspective. DESIPA Working Paper #1. New York: United Nations.

Bell, F. 1989. Application of the wetland valuation theory to Florida fisheries. Florida Sea Grant Program R/C-E-25.

Besnard, W. 1950. Considerações gerais em torno da região lagunar de Cananéia—Iguape. I. *Boletim do Instituto Paulista de Oceanografia* 1(1): 9–26.

Bettencourt, S. U. 1992. Economic valuation of forests and woodlands in sub-Saharan Africa. AFTAG.

Bockstael, N. E., K. E. McConnell, and I. E. Strand. 1989. Measuring the benefits of improvements in water quality: The Chesapeake Bay. *Marine Resource Economics* 6:1–18.

Boto, K. G., and J. S. Bunt. 1981. Tidal export of particulate organic matter from a Northern Australian mangrove system. *Estuarine and Coastal Shelf Science* 13:247–55.

Bovi, M., and A. de Castro. 1993. Assai. In: Clay J., and C. Clement, eds. *Income Generating Forests and Conservation in Amazonia.* Rome: FAO Forestry Department.

Boyle, J. K., W. H. Desvousges, F. R. Johnson, R. W. Dunford, and S. P. Hudson. 1994. An investigation of part-whole biases in contingent valuation studies. *Journal of Environmental Economics and Management* 27: 64–83.

Braat, L. C., S. W. F. Van der Ploeg, and F. Bouma. 1979. Functions of the natural environment, an economic-ecological analysis. I.v.M.-V.U. publication number 79–9. I. S. M. Wereld Natuur Fonds-Netherland.

Brasil. Presidência da República. 1991. O desafio do desenvolvimento sustentável: Relatório do Brasil para a Conferência das Nações Unidas sobre Meio Ambiente e Desenvolvimento. Brasília: CIMA.

Brazil. Various years. Projeto RADAMBRASIL. Rio de Janeiro: Série Levantamento de Recursos Naturais.

Briscoe, J., P. Furtado de Castro, C. Griffin, J. North, and O. Olsen. 1990. Toward equitable and sustainable rural water supplies: A contingent valuation study in Brazil. *The World Bank Economic Review* 4:115–34.

Browder, J. 1988. Public policy and deforestation in the Brazilian Amazon. In: Repetto, R., and M. Gillis, eds. *Public Policy and Misuse of Forest Resources,* 247–79. Cambridge: Cambridge University Press.

Brown, S., A. Gillespie, and A. Lugo. 1991. Biomass estimation methods for tropical forests with applications to forest inventory data. *Forest Science* 35(4):881–89.

Bucher, E. H., and P. C. Huszar. 1995. Critical environmental costs of the Paraguay-Paraná waterway project in South America. *Ecological Economics* 15:3–9.

Bunt, J. S., W. T. Williams, and H. J. Clay. 1982. River water salinity and the distribution of mangrove species along several rivers in North Queensland. *Australian Journal of Botany* 30:401–12.

Camacho, A. S., and T. Bagarinao. 1987. Impact of fishpond development on the mangrove ecosystem in the Philippines. In: UNDP/UNESCO. *Mangroves of Asia and the Pacific: Status and Management.* Technical Report of the UNDP/UNESCO Research and Training Pilot Project on Mangrove Ecosystems in Asia and Pacific (RAS/79/002), pp. 386–406. Quezon City, Philippines: Natural Resources Management Centre and National Mangrove Committee, Ministry of Natural Resources.

Cameron, T. 1988. A new paradigm for valuing non-market goods using referendum data: Maximum likelihood estimation by censored logistic regression. *Journal of Environmental Economics and Management* 15: 355–79.

Cameron, T., and J. Quiggin. 1994. Estimation using CV data from a dichotomous choice with follow-up questionnaire. *Journal of Environmental Economics and Management* 27(3):218–34.

Cardoso, F. H., and E. Faleto. 1969. *Dependency and Development in Latin America.* Berkeley: University of California Press.

CEMIG. 1988. Uso de energia na indústria de ferro-gusa não integrada em Minas Gerais. Belo Horizonte: CEMIG.

Cesario, F. J. 1976. Value of time in recreation benefits studies. *Land Economics* 52(1):32– 41.

Chaves, N. 1985. *Nutrição Básica e Aplicada.* Rio de Janeiro: Editora Guanabara Koogan.

Chichilnisky, G., and G. Heal. In press. *Environmental Markets.* New York: Columbia University Press.

Clawson, M. 1959. Methods of measuring the demand for and value of outdoor recreation. *Resources for the Future* 10:1–36.

Clement, C. 1993. Brazil nut. In: Clay, J., and C. Clement, eds. *Income Generating Forests and Conservation in Amazonia.* Rome: FAO Forestry Department.

Coase, R. 1960. The problem of social cost. *Journal of Law and Economics* 3:1–44.

Cooper, J. C. 1993. Optimal bid selection for dichotomous choice contingent valuation surveys. *Journal of Environmental Economics and Management* 24:25–40.

Costanza, R., ed. 1991. *Ecological Economics: The Science and Management of Sustainability.* New York: Columbia University Press.

Cramer, J. S. 1991. *The Logit Model: An Introduction for Economists.* London: Edward Arnold.

Cummings, R. G., D. Brookshire, and W. Schulze. 1986. *Valuing Environmental Goods: an Assessment of the Contingent Valuation Method.* Savage, MD: Rowman and Littlefield.

Cummings, R., and G. W. Harrison. 1995. The measurement and decomposition of non-use values: A critical review. *Environmental and Resource Economics* 5:225–47.

Daly, H. E. 1992. Allocation, distribution, and scale: Towards an economics that is efficient, just, and sustainable. *Ecological Economics* 6:185–93.

Daly, H. E., and J. B. Cobb, Jr. 1989. *For the Common Good: Redirecting the Economy toward Community, the Environment and a Sustainable Future.* London: Green Print.

Day, J. W. Jr., and A. Yanes-Arancibia. 1987. Coastal lagoons and estuaries as an environment nekton. In: A. Yanez-Arancibia, ed. *Fish Community Ecology in Estuaries and Coastal Lagoons: Towards an Ecosystem Integration,* 17–34. México, DR(R): UNAM.

Desvouges, W. H., R. Johnson, R. Dunford, K. Boyle, S. Hudson, and N. Wilson. 1992. Measuring non-use damages using contingent valuation: An experimental evaluation of accuracy. Raleigh, NC: Research Triangle Institute monograph, 92–101.

Diamond, P. A., and J. Hausman.1994. Contingent valuation: Is some number better than no number? *Journal of Economic Perspectives* 8(4):45–64.

Dieren, W., and M. G. W. Hummelinck. 1979. *Nature's Price: The Economics of Mother Earth.* London: Mariob Boyars.

Drummond, J. A. 1988. O jardim dentro da máquina; Breve história ambiental da Floresta da Tijuca. *Estudos Históricos* (Rio de Janeiro) 1:276–98.

Dugan, P. J. 1990. Wetland Conservation: A Review of Current Issues and Required Action. Gland, Switzerland: IUCN.

Ehrlich, P. R., and A. E. Ehrlich. 1992. The value of biodiversity. *Ambio* 21:219–26.

El Serafy, S. 1988. The proper calculation of income from depletable natural resources. In: Ahmad, J., S. El Serafy, and E. Lutz, eds. *Environmental and Resource Accounting and Their Relevance to the Measurement of Sustainable Development,* 10–18. Washington, DC: The World Bank/UNEP.

EMBRAPA. 1976. Sistema de produção em seringais nativos. Manaus, Brazil: Centro de Pesquisa de Seringueira e Dendê, Circular no. 90.

EMBRAPA/CPATU (Empresa Brasileira de Pesquisa Aropecuária/Centro de Pesquisa Agropecuária do Trópico Umido). 1989. Análise de sistemas de produção pecuária no Município de Paragominas, Estado do Pará. Belém, Pará: EMBRAPA/CPATU, Relatório de Pesquisa.

Environment and Development Commission for Latin America and the Caribbean. 1990. *Nuestra Própria Agenda.* New York: IDB/UNDP.

FAO. 1985. Forest resources of Brazil. Rome: Food and Agriculture Organization of the United Nations. Mimeo.

Farber, S., and R. Costanza. 1987. The economic value of wetland systems. *Journal of Environmental Management* 24:41–51.

FIBGE (Fundação Instituto Brasileiro de Geografia e Estatística). 1981. *Estudo Nacional de Despesa Familiar: Tabelas de Composição de Alimentos.* Rio de Janeiro: FIBGE.

———. 1985. *Censo Agropecuário de 1985.* Rio de Janeiro: FIBGE.

———. 1991. *Censo Demográfico de 1991.* Rio de Janeiro: FIBGE.

Folke, C., K. G. Maler, and C. Perrings. 1992. Biodiversity loss: An introduction. *Ambio* 21:200.

Fundação João Pinheiro. 1988. A siderurgia em Minas Gerais—Análise sócio-econômica do setor guseiro, da produção e comercialização do carvão vegetal. Belo Horizonte. Mimeo.

Funtowicz, S., and J. Ravetz. 1994. The worth of a songbird: Ecological economics as a post-normal science. *Ecological Economics* 10:197–208.

Gentry, B., ed. 1998. *Private Capital Flows and the Environment: Lessons from Latin America.* Cheltenham, UK: Edward Elgar.

Gomes, P. A., and J. B. Oliveira. 1980. Teoria da carbonização da madeira. In: CETEC. *Uso da Madeira para Fins Energéticos.* Belo Horizonte. Série de Publicações Técnicas/SPT-001.

Gonçalves, M. F. R. 1991. Lei orgânica municipal—Sua revisão. *Revista de Administração Municipal* 39(199):22–9.

Gosselink, J. G., E. P. Odum, and R. M. Pope. 1974. The value of the tidal marsh. Baton Rouge, LA: Center for Wetland Resources, Lousiana State University.

Graaf, N. R. de. 1986. *A Silvicultural System for Natural Regeneration of Tropical Rain Forest in Suriname.* Wageningen, The Netherlands: Agricultural University.

Grasso, M. 1994. Avaliação Econômica do Ecossistema Manguezal: Complexo Estuarino-Lagunar de Cananéia, Um Estudo de Caso. Master's thesis. Instituto Oceanográfico, University of São Paulo, SP, Brazil.

Greene, W. H. 1993. *Econometric Analysis.* New York: Macmillan.

Groot, R. S. de. 1992. *Functions of Nature.* Amsterdam: Wolters-Noordhoff.

Hamilton, L. S., and J. A. Dixon. 1989. Mangrove forests: An undervalued resource of the land and the sea. In: Borgese, E. M., N. Ginsburg, and J. R. Morgan, eds. *Ocean Yearbook 8* Chicago: University of Chicago Press.

Hamilton, L. S., and S. C. Snedaker. 1984. *Handbook for Mangrove Area Management.* UNESCO/IUCN.

Hanemann, M. 1994. Valuing the environment through contingent valuation. *Journal of Economic Perspectives* 8(4):19–43.

Hanemann, M., J. Loomis, and B. Kanninen. 1991. Statistical efficiency of double-bounded dichotomous choice valuation. *American Journal of Agricultural Economics* 73:1255– 63.

Hanemann, W. M. 1984. Welfare evaluations in contingent valuation experiments with discrete responses. *American Journal of Agricultural Economics* 66:332–41.

———. 1996. Theory versus data in the contingent value debate. In: Bjornstad, D. J., and J. R. Kahn, eds. *The Contingent Valuation of Environmental Resources: Methodological Issues and Research Needs,* 38–60. Cheltenham, UK: Edward Elgar.

Hanemann, W. M., and B. Kanninen. 1996. The statistical analysis of discrete-response CV data. In: Bateman, I., and K. Willis, eds. *Placing Money Values on the Environment: Theory and Practice of the Contingent Valuation Method in the US, EC and Developing Countries.* Oxford: Oxford University Press.

Hanley, N., C. Spash, and L. Walker. 1995. Problems in valuing the benefits of biodiversity protection. *Environmental and Resource Economics* 5: 249–72.

Hartwick, J. 1977. Intergenerational equity and the investing of rents from exhaustible resources. *American Economic Review* 66:972–4.

Hecht, S. 1985. Environment, development and politics: Capital accumulation and the livestock sector in eastern Amazonia. *World Development* 13:663–84.

Hecht, S., R. Norgaard, and G. Possio. 1988. The economics of cattle ranching in eastern Amazonia. *Interciencia* 13:233–40.

Herz, R. 1990. *Os Manguezais do Brasil.* São Paulo: IOUSP-CIRM.

Hicks, J. 1946. *Value and Capital,* 2nd ed. Oxford: Oxford University Press.

Hoehn, J. P., and A. Randall. 1987. A satisfatory benefit cost indicator from contingent valuation. *Journal of Economics and Management* 14 :226–47.

Holmes, T., and R. Kramer. 1995. An independent sample test of yea-saying and starting point bias in dichotomous-choice contingent valuation. *Journal of Environmental Economics and Management* 29:121–32.

Homma, A. 1989. Extração de recursos naturais renováveis: O caso do extrativismo vegetal na Amazônia. Doctoral dissertation. Department of Rural Economics, Federal University of Viçosa, Minas Gerais, Brazil.

Hueting, R. 1980. *New Scarcity and Economic Growth.* Amsterdam: North Holland.

Hufschmidt, M. M., D. E. James, A. D. Meister, B. Y. Bower, and J. A. Dixon. 1983. *Environment, Natural Systems and Development: an Economic Valuation Guide.* Baltimore, MD: John Hopkins University Press.

IBAMA. 1991. Áreas alteradas por ações antrópicas, no Brasil, por região e por estado. Brasília: Centro de Sensoriamento Remoto.

IBAMA/DIRCOF/DEFIS. 1996 and prior years. Enforcement actions and fines levied. Unpublished data. Brasília: IBAMA.

IBDF. 1983. Inventário Florestal Nacional (Síntese dos Resultados). Brasília, D. F.: Ministério de Agricultura, Instituto Brasileiro de Desenvolvimento Florestal, Depto. de Economia Florestal.

IBGE. 1979. *Matriz de Relações Intersetoriais, Brasil 1975.* Rio de Janeiro: Publicações do Inststituto Brasileiro de Geografia e Estatistica (FIBGE).

———. 1980. *Censo Demográfico.* FIBGE 60(3).

———. 1987. *Contas Nacionais do Brasil 1970–1985.* Rio de Janeiro: IBGE, Depto. de Contas Nacionais.

———. Various years. *Censo Agropecuário.* Rio de Janeiro: FIBGE.

ICED (Inter-Parliamentary Conference on Environment and Development). 1992. Draft final document. Brasília: Inter-Parliamentary Union.

Jansson, A., M. Hammer, C. Folke, and R. Costanza, eds. 1994. *Investing in Natural Capital.* Covelo, CA: Island Press.

Jantzen, J. 1993. *Cost-Effective Pollution Control in Brazil.* The Hague, Netherlands: TME Instituut (Comissioned by the World Bank).

Jardim, M. A., and A. B. Anderson. Undated. Manejo de Populações Nativas do Açaizeiro (*Euterpe oleracea* Mart.) no Estuário Amazônico: Resultados Preliminares Belém: Museu Paraense Emílio Goeldi. Mimeo.

Johansson, P. O. 1987. *The Economic Theory and Measurement of Environmental Benefits.* Cambridge: Cambridge University Press.

Johnson, D. V., and P. K. Nair. 1985. Perennial crop-based agroforestry systems in northeast Brazil. *Agroforestry Systems* 2:281–92.

Jordan, J., and A. Elnagheeb. 1994. Consequences of using different question formats in contingent valuation: A Monte Carlo study. *Land Economics* 70(1):97–110 .

Judge, G. G., et al. 1985. *The Theory and Practice of Econometrics.* New York: Wiley.

Kaneham, D., and L. J. Knestch. 1992. Valuing public goods: The purchase of moral satisfaction. *Journal of Environmental Economics and Management* 22:57–70.

Kanninen, B. 1995. Bias in discrete response contingent valuation. *Journal of Environmental Economics and Management* 28:114–25.

King, D. 1994. Justifying sustainability: Some basics of applied ecological economics. In: Jansson, A., M. Hammer, C. Folke, and R. Costanza, eds. *Investing in Natural Capital.* Washington, DC: Island Press.

Kreimer, A., T. Lobo, B. Menezes, M. Munasinghe, R. Parker, and M. Preece. 1993. Rio de Janeiro—The search for sustainability. In: World Bank, *Towards a Sustainable Urban Environment: The Rio de Janeiro Study.* Washington, DC: International Bank for Reconstruction and Development.

Kristrom, B. 1990. A non-parametric approach to the estimation of welfare measures in discrete response valuation studies. *Land Economics* 66(2): 135–9.

Kuznets, S. 1941. *National Income and Its Composition, 1931–1938.* New York: National Bureau of Economic Research, vol. 1.

Lake, E. B., and A. M. Shady. 1993. Erosion reaches crisis proportions. *Agricultural Engineering* Nov:8–13.

Lal, P. N. 1990. Conservation or conversion of mangroves in Fiji. Hawaii: Occasional papers of the East-West Environmental and Policy Institute, paper no. 11.

Lee, E. T. 1992. *Statistical Methods for Survival Data Analysis.* New York: Wiley.

Li, C. Z., and L. Mattsson. 1995. Discrete choice under preference uncertainty: An improved structural model for contingent valuation. *Journal of Environmental Economics and Management* 28:256–69.

Lima, W. P. 1993. *Impacto Ambiental do Eucalipto.* São Paulo: Universidade de São Paulo.

Loomis, J. B. 1998. Contingent valuation using dichotomous choice models. *Journal of Leisure Research* 20(1):46–56.

Loomis, J., and P. duVair. 1993. Evaluating the effect of alternative risk communication devices on willingness to pay: Results from a dichotomous choice contingent valuation experiment. *Land Economics* 69(3):287–98.

Lopes, I., G. Bastos, D. Biller, and M. Bale, eds. 1996. *Gestão Ambiental no Brasil, Experiência e Successo.* Rio de Janeiro: Fundação Getúlio Vargas/ World Bank.

Lutz, E., ed. 1993. Toward improved accounting for the environment. *A UNSTAT-World Bank Symposium.* Washington, DC: The World Bank.

Lynne, G. D., P. Conroy, and F. J. Prochaska. 1981. Economic valuation of marsh areas for marine production process. *Journal of Environmental Economics and Management* 8(2):175–86.

Maciel, N. C. 1986. Desarraigamento de Manguezal, Através de Barragem de Rios e Gamboas com Aterro para Implantação de Salina, no Município de Galinhos, Rio Grande do Norte. Parecer Técnico. Rio de Janeiro: Fundação Estadual de Engenharia do Meio Ambiente.

Macmillan, D., N. Hanley, and S. Buckland. 1996. Valuing biodiversity losses due to acid deposition: A contingent valuation study of uncertain environmental gains. *Scottish Journal of Political Economy* 43(5):519–33.

Macnae, W. 1974. Mangrove forests and fisheries. FAO/UNDP Indian Ocean Programme. IOFC/DEV/7434.

Magalhães Gomes, F. 1983. *História da Siderurgia no Brasil.* Belo Horizonte: Itatiaia.

Magalhães, J. L. 1993. Futuro do carvão vegetal no contexto nacional e no exterior. *Anais do I Simpósio Brasileiro de Pesquisa Florestal.* Belo Horizonte: Sociedade de Investigações Florestais.

Martines, J., et al. 1991. *Health Sector Priorities Review.* Washington, DC: Population and Human Resources Department, The World Bank.

Martinez-Alier, J. 1994. Distributional obstacles to international environmental policy (the failures at Rio and prospects after Rio). In: Jansson, A., M. Hammer, C. Folke, and R. Costanza, eds. *Investing in Natural Capital.* Washington, DC: Island Press.

———. In press. Ecological economics as human ecology. In: Costanza, R., and S. Tognetti, eds. *Ecological Economics and Integrated Assessment: A Participatory Framework for Including Equity, Efficiency and Scale in Decisionmaking.* Sponsored by SCOPE/UNEP.

Martosubroto, P. D., and N. Naamin. 1977. Relationship between tidal forests (mangroves) and commercial shrimp production in Indonesia. *Market Research in Indonesia* 18:81–6.

Mattos, M., and C. Uhl. 1994. Economic and ecological perspectives on ranching in the Eastern Amazon. *World Development* 22:145–58.

Max-Neef, M. 1992. Development and human needs. In: Ekins, P., and M. Max-Neef, eds. *Real-life Economics: Understanding Wealth Creation,* 197–213. London: Routledge.

May, P. 1993. Estimativas de Contas Ambientais; Perdas Ambientais Devidos ao Desmatamento: Relatório Final. Rio de Janeiro: IPEA-Rio. Mimeo.

———. 1994. Measuring sustainability: Forest values and agropastoral expansion in Brazil. In: Duijnhouwer, J., G. J. Van der Meer, and H. Verbrugger, eds. *Natural Resource Conservation and Resource Use,* 139–163. The Netherlands: RMNO.

———. 1995. A Survey of Environmentally Friendly Products of Brazil. Geneva: UNCTAD. Mimeo.

———. 1996. Limits to Brazilian deforestation: An exercise in environmental accounts estimation. Presented at a *Workshop on Valuation Methods for Green National Accounting: A Practical Guide,* The World Bank, 20–22 March.

———., ed. 1995. *Economia Ecológica: Aplicações no Brasil.* Rio de Janeiro: Editora Campus.

May, P., A. B. Anderson, J. M. Frazão, and M. J. Balick. 1985. Babassu palm in the agroforestry systems in Brazil's mid-north region. *Agroforestry Systems* 3:275–95.

May, P., and M. Pastuk. 1996. Valuing social sustainability: Environmental recuperation on *favela* hillsides in Rio de Janeiro. In: Segura Bonilla, O., R. Costanza, and J. Martinez-Alier, eds. *Getting Down to Earth: Practical Applications of Ecological Economics.* Washington, DC: Island Press.

May, P., and O. Segura Bonilla. 1997. The environmental effects of agricultural trade liberalization in Latin America: An interpretation. *Ecological Economics* 21:8–15.

McConnell, K. E. 1975. Some problems in estimating the demand for outdoor recreation. *American Journal of Agricultural Economics* 56:330–4.

———. 1988. Introducing referendum models. Paper presented at IDB Workshop on Valuation Techniques in Project Analysis, November 21.

———. 1990. Models for referendum data: The structure of discrete choice models for contingent valuation. *Journal of Environmental Economics and Management* 18:19– 34.

McLelland, G., W. Schulze, W. D. Lazo, J. Doyle, R. Sleve, and J. Irwin. 1992. *Measures for measuring non-use values: A contingent valuation study of groundwater cleanup.* University of Colorado, Center for Economics Analysis.

Medeiros, J. X. 1993. Suprimento energético de carvão vegetal no Brasil: Aspectos técnicos, econômicos e ambientais. *Anais do VI Congresso Brasileiro de Energia,* Rio de Janeiro. 1:107–12.

———. 1995. Energia renovável na siderurgia: Análise sócio-econômica e ambiental da produção de carvão vegetal para os altos fornos de Minas Gerais (no início da década de 1990). Doctoral dissertation. UNICAMP, Campinas.

Mendelsohn, R. 1987. Modeling the demand for outdoor recreation. *Water Resources Research* 23(5):961–7.

MIC/STI. 1982. Levantamento de áreas de ocorrência e produtividade do Babaçu. Brasília. Ministério de Indústria e do Comércio.

Mitchell, H. S., H. J. Rynberger, L. Anderson, and M. V. Dibble. 1978. *Nutrição.* Rio de Janeiro: Editora Interamericana.

Mitchell, R. C., and R. Carson. 1989. *Using Surveys to Value Public Goods: The Contingent Valuation Method.* Washington, DC: Resources for the Future.

Montague, C. L., A. V. Zale, and H. F. Percival. 1987. Ecological effects of coastal marsh impoundments: A review. *Environmental Managagement* 11(6):753–6.

Moran, E. F. 1981. *Developing the Amazon.* Bloomington, IN: Indiania University Press.

Mori, S. A., and G. T. Prance. 1990. Taxonomy, ecology and economic botany of the Brazil Nut (*Bertholletia excelsa* Humb. and Bonpl.: Lecythidaceae). *Advances in Economic Botany* 8:130–50.

Moulton, T. 1991. Putting a value on the mangrove/estuarine system of Cananéia/Iguape. Mimeo. Unpublished manuscript.

Mourão, F. A. A. 1971. Os pescadores do litoral sul do Estado de São Paulo. Doctoral thesis, Faculdade de Filosofia, Ciências e Letras. Universidade de São Paulo, SP, Brazil.

Munasinghe, M., B. Menezes, and M. Preece. 1990. Case study: Rio flood reconstruction and prevention project. In: Kramer, A., and M. Munasinghe, eds. *Managing Natural Disasters and the Environment,* 28–31. Washington, DC: World Bank Environment Department.

National Academy of Sciences. 1977. Methane generation from human, animal, and agricultural wastes. Panel on Methane Generation of the Advisory Committee on Tecnology Innovation. Washington, DC: NAS/AID.

National Oceanic and Atmospheric Administration (NOAA). 1993. Report of the NOAA panel on contingent valuation. *Federal Register* 58(10): 4602–14.

Nepstad, D.C. 1989. Forest regrowth in abandoned pastures of eastern Amazônia: Limitations to tree seedling survival and growth. PhD dissertation. New Haven, CT: Yale University.

Nordhaus, W., and J. Tobin. 1992. Is growth obsolete? In: *Economic Growth.* New York: Columbia University Press, National Bureau of Economic Research General Series #96e.

Norgaard, R. 1994. *Development Betrayed; The End of Progress and a Coevolutionary Revisioning of the Future.* London: Routledge.

Odum, H. Y. 1971. *Environment, Power and Society.* New York: Wiley.

Odum, W. E., J. S. Fisher, and J. C. Pickral. 1979. Factors controlling the flux of particulate organic carbon from estuarine wetlands. In: Livingston, R. J., ed. *Ecological Processes in Coastal and Marine Systems, Marine Science Series,* 10:69–80. New York: Plenum Press.

Odum, W. E., and E. J. Heald. 1972. Trophic analysis of an estuarine mangrove community. *Bulletin of Marine Science* 22:671–738.

Ostro, B. 1992. The health effects of air pollution: A methodology with applications to Jakarta. Washington, DC: The World Bank. Mimeo.

Park, T., J. Loomis, and M. Creel. 1991. Confidence intervals for evaluating benefits estimates from dichotomous choice contingent valuation studies. *Land Economics* 67(1):64–73.

Pastuk, M. 1995. Urban and peri-urban forestry in the Rio de Janeiro metropolitan area. Rome: Report to FAO Forestry Department.

Pauly, D., and J. Ingles. 1986. The relationship between shrimp yields and intertidal vegetation (mangrove) areas: A reassessment. In IOC/FAO Workshop on Recruitment in Tropical Coastal Demersal Communities, submitted papers, 227–84. Paris: UNESCO.

Pearce, D., and G. Atkinson. 1992. Are national economies sustainable? Measuring sustainable development. London and East Anglia: Centre for Social and Economic Research on the Global Environment (CSERGE). Working Paper GEC, 92–111.

Pearce, D. W. 1976. *Environmental Economics.* London: Longman.

Pearce, D. W., and D. Moran. 1994. *The Economic Value of Biodiversity.* London: Earthscan.

Perrings, C. 1995. Economic values of biodiversity. In: *Global Biodiversity Assessment.* Cambridge, UK: United Nations Environmental Program and University of Cambridge Press.

Pool, D. J., S. C., Snedaker, and A. E. Lugo. 1977. Structure of mangrove forests in Florida, Puerto Rico and Costa Rica. *Biotropica* 9:195–212.

Portney, P. 1994. The contingent valuation debate: Why economists should care. *Journal of Economic Perspectives* 8(4):3–17.

Pupo, N. I. H. 1981. *Manual de pastagens e forrageiras: formação, conservação e utilização.* Campinas: Instituto Campineiro de Ensino Agrícola.

Ramdial, B. S. 1980. The social and economic importance of the Caroni mangrove swamp forests. Commonwealth Forestry Conference.

Randall, A., and W. Kriesel. 1990. Evaluating national policy proposals by contingent valuation. In: Johnson, R., and G. Johnson, eds. *Economic Valuation of Natural Resources: Issues, Theory and Applications.* Boulder, CO: Westview.

Reis, E., and S. Margulis. 1991. Perspectivas econômicas do desflorestamento da Amazônia. Brasília: IPEA/Ministério da Economia, Fazenda e Planejamento, Texto para Discussão no. 215.

Renesto, O. V., and L. F. Vieira. 1977. *Analise Econômica da Produção e Processamento do Palmito em Conserva nas Regiões Sudeste e Sul do Brasil.* Campinas, Brazil: ITAL Estudos Econômicos, Alimentos Processados 6.

Repetto, R., W. Magrath, M. Wells, C. Beer, and F. Rossini. 1989. *Wasting Assets: Natural Resources in the National Income Accounts.* Washington, DC: World Resources Institute.

Ribeiro, S. I. 1989. Citrus: Informações básicas para seu cultivo no Estado do Pará. Belém, Pará: EMBRAPA/UEPAE.

Robertson, A. I., and D. M. Alongi. 1992. *Tropical Mangrove Ecosystem.* Washington, DC: American Geophysical Union, Coastal and Estuarine Studies #41.

Saenger, P., E. J. Hergel, and J. D. S. Davie. 1983. *Global Status of Mangrove Ecosystems.* Gland, Switzerland: International Union for Conservation of Nature and Natural Resources, Commission on Ecology papers.

Sánchez, P. A. 1976. *Properties and Management of Soils in the Tropics.* New York: Wiley-Interscience.

Schneider, R. 1993. The potential for trade with the Amazon in greenhouse gas reduction. Washington, DC: The World Bank, LATEN Dissemination Note no. 2.

Serôa da Motta, R. 1991. Mecanismos de mercado na política ambiental brasileira. In: *Perspectiva da Economia Brasileira—1992.* Rio de Janeiro: IPEA.

Serôa da Motta, R., ed. 1995. *Contabilidade Ambiental: Teoria, Metodologia e Estudos de Casos no Brasil.* Rio de Janeiro: IPEA.

Serôa da Motta, R., Fernandes Mendes, A. P., Mendes, F. E., and Young, C. E. 1994. Perdas e serviços ambientais do recurso água para uso doméstico. *Pesquisa e Planejamento Econômico* 24:1.

Serôa da Motta, R., and P. May. 1992. Loss in forest resource values due to agricultural land conversion in Brazil. Rio de Janeiro: IPEA-Rio. Texto para Discussão 248.

———. 1996. Natural resource exploitation and sustainable development in Brazil. In: May, P., and R. Serôa da Motta, eds. *Pricing the Planet: Economic Analysis for Sustainable Development,* 197–208. New York: Columbia University Press.

Serôa da Motta, R., and A. P. Mendes. 1993. Health costs associated to air pollution in Brazil. Rio de Janeiro: IPEA. Mimeo.

Serôa da Motta, R., and F. E. Mendes. 1995. Perdas ambientais associadas à poluição de recursos hídricos. In: Serôa da Motta, R., ed. *Contabilidade Ambiental: Teoria, Metodologia e Estudos de Casos no Brasil.* Rio de Janeiro: IPEA.

Serôa da Motta, R., and E. J. Reis. 1994. The application of economic instruments in environmental policy: The Brazilian case. Paper presented at the *Workshop on the Use of Economic Policy Instruments for Environmental Management,* OECD/UNEP, Paris, 26–27 May.

Serôa da Motta, R., and C. E. F. Young. 1991. Recursos naturais e contabilidade social: A renda sustentável da extração mineral no Brasil. Rio de Janeiro: IPEA-Rio. Texto para Discussão 231.

Serrão, A., and I. Falesi. 1977. Pastagens do trópico umido Brasileiro. Belém, Pará: EMBRAPA/CPATU.

Sherman, P. B., and J. A. Dixon. 1990. *Economics of Protected Areas: A New Look at Benefits and Costs.* Washington, DC: East-West Center/Island Press.

Siqueira, J., and D. Pierin. 1990. A atividade florestal como um dos instrumentos de desenvolvimento do Brasil. *Anais do 6o Congresso Florestal Brasileiro.* 1:15–8. São Paulo: SBS/SBEF.

SMA. 1990. Macrozoneamento do complexo estuarino-lagunar de Iguape-Cananéia. Plano de gerenciamento costeiro. São Paulo, Secretaria do Meio Ambiente, Série documentos.

Smith, V. K. 1983. Taking stock of progress with travel cost recreation demand methods: Theory and implementation. *Marine Resource Economics* 6:279–310.

———. 1992. Arbitrary values, good causes, and premature verdicts. *Journal of Environmental Economics and Management* 22:71–89.

Solárzano, R., et al. 1992. *Accounts Overdue: Natural Resource Depletion in Costa Rica.* Washington, DC: World Resources Institute.

SOS Mata Atlântica. 1992. *Atlas dos Remanescentes da Mata Atlântica.* São Paulo.

Squire, L., and H. Van der Tak. 1975. *Economic Analysis of Projects.* Baltimore, MD: Johns Hopkins University Press.

Staples, D. J., D. J. Vance, and D. S. Heales. 1985. Habitat requirements of juvenile *Penaied* prawns and their relationship to offshore fisheries. In: Rothlisberg, P. C., B. J. Hill, and D. J. Staples, eds. *Second Australian National Prawn Seminar,* 47–54. Cleveland, Australia.

Stout, B. A. 1980. *Energia para la Agricultura Mundial.* Rome, Italy: FAO.

Subler, S. E. 1993. Mechanisms of nutrient retention and recycling in a chro-

nosequence of Amazonian agroforestry systems: Comparisons with natural forest ecosystems. PhD dissertation. Pennsylvania State University.
Thibau, C. E. 1972. Supply of charcoal to the Brazilian pig-iron industry. Brasília: IBDF/Ministério da Agricultura. Mimeo.
Tietenberg, T. H. 1990. Economic instruments for environmental regulation. *Oxford Review of Economic Policy* 6:1.
Toniolo, A., and C. Uhl. 1995. Economic and ecological perspectives on agriculture in the Eastern Amazon. *World Development* 23:959–73.
Turner, R. E. 1977. Intertidal vegetation and commercial yields of *Penaeid* shrimp. *Transactions of the American Fishery Society* 106:411–6.
Turner, R. E., and D. F. Boesh. 1989. Aquatic animal production and wetland relationships: Insights gleaned following wetland loss or gain. In: Hook, D. D., et al., eds. *The Ecology and Management of Wetlands,* vol 1: Ecology of Wetlands. Portland, Oregon: Croom Helm.
Turner, R. K., et al. 1992. Sea level rise and coastal wetlands in the UK: Mitigation strategies for sustainable management. In: Jansson, A., M. Hammer, C. Folke, and R. Costanza, eds. *Investing in Natural Capital,* 266–90. Covelo, CA: Island Press.
Twilley, R. R. 1988. Coupling of mangroves to the productivity of estuarine and coastal waters. *Lectures on North Coastal Estuarine Studies* 22: 155–80.
Uhl, C., R. Buschbacher, and E. A. S. Serrão. 1988. Abandoned pastures in Eastern Amazonia. Patterns of plant succession. *Journal of Ecology* 76: 663–81.
U.N. Development Program (UNDP). 1995. *Human Development Report 1995,* New York.
UNDP/IBAM. 1993. Programa de preservação e reflorestamento em regiões de Baixa Renda—Projeto Mutirão/Reflorestamento. Rio de Janeiro: Megacities Project.
van Tongeren, J., S. Schweinfest, E. Lutz, M. Gomez Luna, and G. Martin. 1991. *Integrated Environmental and Economic Accounts: A Case Study for Mexico.* Washington, DC: The World Bank, Environmental Working Paper no. 50.
Veiga, J. B. 1986. Associação de culturas de subsistencia com forrageiras na renovação de pastagens degradadas em areas de floresta. *Anais do Primeiro Simpósio do Trópico Úmido.* Brasília: Departamento de Difusão de Tecnologia, EMBRAPA.
Veloso, H. P., and L. Goes Filho. 1982. *A Vegetação do Brasil.* Rio de Janeiro: FIBGE.
Veríssimo, A., P. Barreto, M. Mattos, R. Tarifa, and C. Uhl. 1992. Logging impacts and prospects for sustainable forest management in an old Amazonian frontier: The case of Paragominas. *Forest Ecology and Management* 55:169–99.
Watrin, O. S., and A. M. Rocha. 1992. Levantamento da vegetação natural

e do uso da terra no município de Paragominas PA utilizando imagens TM/LANDSAT. Belém, Pará: EMBRAPA/CPATU, Boletim de Pesquisa, no. 124.

Whittington, D., J. Briscoe, X. Mu, and W. Barro. 1991a. Estimating the willingness to pay for water services in developing countries: A case study of the use of contingent valuation surveys in southern Haiti. *Economic Development and Cultural Change* 38(2):293–311.

Whittington, D., G. Cassidy, D. Amaral, E. McClelland, H. Wang, and C. Poulos. 1994. *The Economic Value of Improving the Environmental Quality of Galveston Bay.* Department of Environmental Sciences and Engineering, University of North Carolina, Galveston Bay Estuary Programme, Chapel Hill, NC.

Whittington, D., V. Kerry Smith, A. Okorafor, A. Okore, J. L. Lui, and A. McPhail. 1991b. Giving respondents time to think in contingent valuation studies: A developing country application. *Journal of Environmental Economics and Management* 21:205–25.

Willis, K., G. Garrod, and C. Saunder. 1995. Benefits of environmentally sensitive area policy in England: A contingent valuation assessment. *Journal of Environmental Management* 44:105–25.

Winograd, M. 1995. Indicadores ambientales para Latinoamérica y el Caribe: Hacia la sustentabilidad en el uso de tierras. San José, Costa Rica: International Institute for Agronomic Science (IICA).

World Bank. 1992. *Development and the Environment (1992 World Development Report).* Washington, DC: International Bank for Reconstruction and Development.

Index

Head note: Except for specified other nations or states, all subject headings refer to their application, existence, or practice in Brazil. *Italic page numbers* followed by *italic letters e, f, or t* refer, respectively, to an *e*quation, *f*igure, or *t*able. The letter *n* following page numbers refers to chapter notes, e.g., 26n.1 is note 1 on page 26.